A CO-REQUISITE WORKBOOK FOR
MATH FOR GENERAL STUDIES

PELLISSIPPI STATE COMMUNITY COLLEGE
MATHEMATICS FACULTY

Kendall Hunt
publishing company

Cover image © Shutterstock, Inc.

Kendall Hunt
publishing company

www.kendallhunt.com
Send all inquiries to:
4050 Westmark Drive
Dubuque, IA 52004-1840

Copyright © 2017 by Kendall Hunt Publishing Company

ISBN 978-1-5249-0462-3

Printed in the United States of America

TABLE OF CONTENTS

FOREWORD

Over the past few years, the *Tennessee Board of Regents,* which governs community colleges throughout the state of Tennessee, has chosen to eliminate stand-alone learning support classes that were designed to remediate skills in mathematics for students whose skills were below the college entry level. To reach this goal, the community colleges were directed to create co-requisite classes that help students overcome their deficits while they are concurrently taking their entry-level math courses. Pellissippi State Community College's math department has been at the leading edge of this initiative.

Remedial mathematics is decidedly "algebra-heavy," with topics such as polynomial multiplication and factoring that are of little use to a non-STEM student. Many majors at community colleges require only a general studies math course, which helps students to be mathematically literate, but does not cover all areas of algebra. This has led to the creation of a specialized curriculum for the co-requisite course. Specifically, the context of all the content in this workbook is geared directly to the needs of a Math for General Studies course. It is designed to provide *just-in-time* remediation and is directly tied to concepts in the college-level course. In this way, the authors felt that students would see this as a direct aid to their success in the college-level course, rather than simply another math course.

The workbook structure is arranged to support the customary coverage of a survey of math course and could support a variety of textbooks. Commonly misunderstood concepts, such as the order of operations, are encountered in multiple places from simple formulas, such as the mean, to complex formulas, such as those found in finance. The title heading of each unit and lesson should allow professors to pick and choose workbook sections that might be of help to individual students.

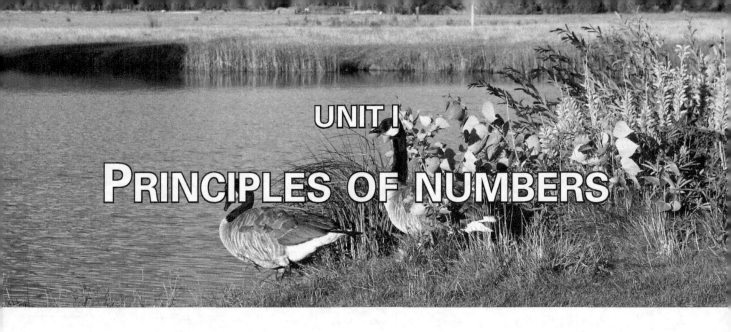

UNIT I
PRINCIPLES OF NUMBERS

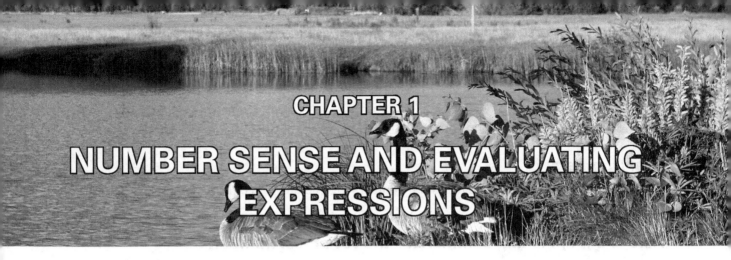

CHAPTER 1

NUMBER SENSE AND EVALUATING EXPRESSIONS

1.1 Identifying Sets of Numbers and Absolute Value

There are a number of different ways to classify numbers. This section will introduce the most common classifications of real numbers so you can explore, apply, and master this concept. Specifically, we will investigate real number classification as it applies to commonly encountered numbers.

Natural Numbers:

The natural numbers start with 1 and include the numbers $\{1, 2, 3, 4 \ldots\}$ and so on. The ellipsis (. . .) symbol indicates that you continue in the pattern established. These are the "counting" numbers and are all positive. Zero is not considered a "natural number."

Whole Numbers:

The whole numbers start with zero and are composed of the numbers $\{0, 1, 2, 3, 4 \ldots\}$. The whole numbers are all the natural numbers plus zero.

Integers:

The integers are the whole numbers and their opposites: the positive whole numbers, the negative whole numbers, and zero. Integers can be shown as the set $\{\ldots -3, -2, -1, 0, 1, 2, 3 \ldots\}$.

Rational Numbers:

The rational numbers are made up of all the integers plus fractions that can be represented as a ratio of two whole numbers. This includes terminating decimals, which end naturally without rounding, and repeating decimals, which have a pattern of numbers that repeats infinitely but does not end. For example, 0.357 is a terminating decimal, and 0.818181818181 . . . is a repeating decimal. The repeating decimal can also be written $0.\overline{81}$ to show that the sequence repeats itself.

Irrational Numbers:

An irrational number is a number with a decimal that neither terminates nor repeats. For example, $\sqrt{2} \approx 1.4142135624 \ldots$ is a decimal that does not repeat itself, but that continues infinitely. Any square root of a number that is not a perfect square is an irrational number. The value of $\pi \approx 3.1415926536 \ldots$ is also classified as an irrational number. We often approximate these irrational numbers with terminating

3

decimals. These approximations, such as, $\pi \approx 3.14$ become so commonly used that we often think of them as being "equal to" the value of the irrational number, but they are not.

Real Numbers:

The combined set of rational and irrational numbers together are called the real numbers. This is the set represented by a "number line."

Here is a graphic that summarizes the set of real numbers:

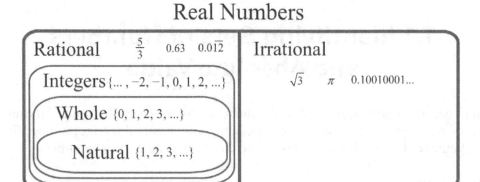

EXAMPLE 1

Classify the numbers $3, \sqrt{5}, \dfrac{2}{3}$ by placing an X in any appropriate category.

	Natural Number	Whole Number	Integer	Rational Number	Irrational Number	Real Number
3	X	X	X	X		X
$\sqrt{5}$					X	X
$\dfrac{2}{3}$				X		X

- 3 is an integer as well as a rational number (3/1 is the fractional equivalent) and a real number since all rational numbers are real numbers.
- $\sqrt{5}$ is an irrational number since it is the square root and 5 is not a perfect square. It is also a real number since all irrationals are real numbers.
- $\dfrac{2}{3}$ is a rational number because it is a ratio or fraction of two whole numbers and can also be written as a repeating decimal, $0.\overline{6}$. It is also a real number.

Number Lines:

Number lines are a visual way of representing all real numbers. An understanding of number lines is necessary to see the relationships in inequality statements as they relate to probability.

There are a few basic ideas underpinning an understanding of number line order:

1. Negative numbers are to the left of zero and positive numbers are to the right of zero.
2. The number line extends infinitely to both the left and the right.
3. As you go from left to right, the numbers get larger or increase. This may be confusing with negative numbers because it means that -1 is larger than -2 for example.

Here is a blank number line:

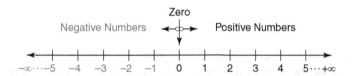

Keeping these ideas in mind, placing numbers correctly on the number line requires that we be able to compare the numbers and determine their relative order from smallest to largest. Perhaps, the easiest way to do this is to express all the numbers in decimal form, being sure that each comparison pair has the same number of decimal places. Then we can place the numbers on the number line.

EXAMPLE 2

Place the numbers in the following data set on the number line:

$$-2.4, 0.367, 4/5, 0.1, 1.6, -2/3$$

Step 1: Express the numbers in decimal form with the same number of decimal places. In this case, each number should be shown with three decimal places since there is one number in the list with three decimal places.

Using a calculator to find the equivalents when needed:

$$-2.4 = -2.400$$
$$\frac{4}{5} = 0.800$$
$$0.1 = 0.100$$
$$1.6 = 1.600$$
$$-\frac{2}{3} = -0.667 \ (rounded)$$

Step 2: Sort the numbers from smallest to largest: $-2.400, -0.667, 0.100, 0.367, 0.800, 1.600$

Step 3: Now place these numbers on the number line:

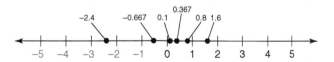

We can also state explicitly that the following inequality is true:

$$-2.4 < -0.667 < 0.1 < 0.367 < 0.8 < 1.6$$

EXAMPLE 3

Graph 3, $\dfrac{2}{3}$, and $\sqrt{5}$ on a number line:

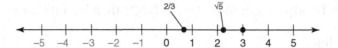

Absolute Value:

On a number line, the distance from zero is called the absolute value. All real numbers have an absolute value that can be determined by examining how many units from zero a number is. For example, the number 3 is three units to the right of zero, while the number -3 is three units to the left of zero. The numbers 3 and -3 have the same absolute value because they are the same distance from zero.

1.1 PRACTICE PROBLEMS

1. Classify each number by checking all of the appropriate boxes:

	Natural Number	Whole Number	Integer	Rational Number	Irrational Number	Real Number
−2						
$\frac{7}{3}$						
$\sqrt{4}$						
5.3						
0						
$\sqrt{7}$						
24						

2. Graph the following numbers on the number line provided: $1.3, -4, \sqrt{8}, \frac{3}{4}$.

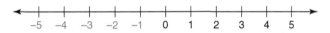

3. Juliette measures the temperature inside a barn at four hour intervals for a day. This is the data set she obtains (in degrees Fahrenheit): three below zero, zero, five, four, one, and one below zero.

   a. List these readings numerically:

   b. Graph these readings on the number line below:

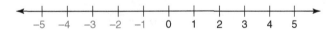

4. What number is the "opposite" of the number 2?

5. Graph both the number 2 and its opposite on the number line below:

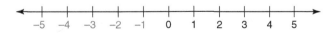

6. What is the absolute value of −5?

7. Why do −10 and 10 have the same absolute value?

8. Order the following numbers 5, −3, 2.5, −1 from largest to smallest.

1.2 Order of Operations to Simplify Expressions

Order of Operations Review:

You will frequently use the order of operations to perform algebraic computations. The order of operations are as follows:

Parentheses | Exponents | Multiplication or Division | Addition or Subtraction

1. Perform the operations inside parentheses and other grouping symbols such as brackets first.
2. Evaluate exponents.
3. Do multiplication and division, from left to right.
4. Do addition and subtraction, from left to right.

The acronym PEMDAS is used as a mnemonic device to help remember the order of operations.

EXAMPLE 1

Evaluate $6 \cdot (2+8)$.

Step 1: We note that the order of operations requires us to solve inside the parenthesis first.

$$6 \cdot (2+8)$$
$$= 6 \cdot 10$$

Step 2: Since there are no exponents in this expression, next we do the multiplication.

$$6 \cdot 10$$
$$= 60$$

EXAMPLE 2

Evaluate $6.3 + 2 \cdot 1.2$.

Step 1: There are no parentheses or exponents in this expression. The multiplication must be done first.

$$6.3 + 2 \cdot 1.2 = 6.3 + 2.4$$

Step 2: Now do the indicated addition.

$$6.3 + 2.4 = 8.7$$

EXAMPLE 3

Evaluate $(3+5)^2 - 4 \cdot 6$.

Step 1: The parenthesis must be done first.

$$(3+5)^2 - 4 \cdot 6 = 8^2 - 4 \cdot 6$$

Step 2: The exponent must be done next.

$$8^2 - 4 \cdot 6 = 64 - 4 \cdot 6$$

Step 3: Now we must do the multiplication.

$$64 - 4 \cdot 6 = 64 - 24.$$

Step 4: Finally, subtract the two values.

$$64 - 24 = 40$$

EXAMPLE 4

Evaluate $\dfrac{5 - 3 \cdot 3}{2}$.

Step 1: In using the order of operations, we have an "implied" parenthesis in both the numerator and the denominator of any fraction. Therefore, in this problem, we do the numerator operations first.

$$\frac{5 - 3 \cdot 3}{2} = \frac{(5 - 3 \cdot 3)}{2} = \frac{(5 - 9)}{2} = \frac{-4}{2}$$

Note that in doing the operations in the numerator, we have continued to apply the order of operations by doing the multiplication before the subtraction.

Step 2: Now that the numerator operations have been performed, we can do the division called for by the fractional form.

$$\frac{-4}{2} = -4 \div 2 = -2$$

1.2 PRACTICE PROBLEMS

Use order of operations to simplify the following problems. Show all of your work.

1. $12 + 2 \cdot 5$

2. $5^2 + 7 \cdot 9 \div 3 - 2$

3. $\dfrac{20 - 12}{4}$

4. $\dfrac{7 + 4 + 9 + 2 + 3}{5}$

5. $25 + 4 \cdot 3$

6. $3^2 - 12 \div 2^2 \cdot 6 - 4$

7. $\dfrac{20 + 15 \div (-3)}{5^2}$

8. $15 \div \dfrac{5^2}{9 - 4} - (-2)$

9. $6(3 + 2)$

10. $24 \div 3 \cdot 2$

1.3 Evaluating Algebraic Expressions

One crucial skill needed to be successful in mathematics is the ability to correctly connect the numbers in a word problem with the symbolic concept that they represent. In addition, you need to then be able to substitute properly into the formula given and use the order of operations correctly to solve.

Consider the following examples.

EXAMPLE 1

Evaluate the algebraic expression

$$3x - 5y + 6$$

when $x = 3$ and $y = 2$.

> **Step 1:** We must first substitute the values given for x and y into the expression.
>
> $$3(3) - 5(2) + 6$$
>
> Note that the term $3x$ means $3 \cdot x$. Whenever numbers are shown next to variables, multiplication is indicated.
>
> **Step 2:** Now follow the order of operations for evaluating the expression. The multiplications must be done first.
>
> $$3(3) - 5(2) + 6 = 9 - 10 + 6$$
>
> **Step 3:** Complete the evaluation by performing the addition and subtraction from left to right.
>
> $$9 - 10 + 6 = -1 + 6 = 5$$

EXAMPLE 2

Evaluate the algebraic expression

$$\frac{3 - x}{x + y}$$

when x is 2 and y is 4.

> **Step 1:** There is an implied parenthesis around both the numerator and the denominator.
>
> $$\frac{3 - x}{x + y} = \frac{(3 - x)}{(x + y)}$$
>
> **Step 2:** Substitute the known values of x and y into the expression.
>
> $$\frac{3 - x}{x + y} = \frac{(3 - 2)}{(2 + 4)}$$
>
> **Step 3:** Perform the operations in the parentheses.
>
> $$\frac{3 - x}{x + y} = \frac{(3 - 2)}{(2 + 4)} = \frac{1}{6}$$

EXAMPLE 3

Evaluate the algebraic expression

$$2r - 3c^2 + (r + c)$$

when $r = 5$ and $c = 2$.

Step 1: Substitute the known values of r and c into the expression.

$$2r - 3c^2 + (r + c) = 2(5) - 3(2)^2 + (5 + 2)$$

Step 2: The parenthesis operation must be done first.

$$2(5) - 3(2)^2 + (5 + 2) = 2 \cdot 5 - 3(2)^2 + 7$$

Step 3: Next, you must evaluate the exponent.

$$2 \cdot 5 - 3(2)^2 + 7 = 2 \cdot 5 - 3 \cdot 4 + 7$$

Step 4: Continuing to follow the order of operations, the multiplications are performed.

$$2 \cdot 5 - 3 \cdot 4 + 7 = 10 - 12 + 7$$

Step 5: Finally, perform the addition and subtraction from left to right.

$$10 - 12 + 7 = -2 + 7 = 5$$

1.3 PRACTICE PROBLEMS

1. Evaluate the algebraic expression for $a = 2$

$$-3 + a$$

2. Evaluate the algebraic expression for $x = 3$ and $y = -2$

$$2(x - y)$$

3. Evaluate the algebraic expression for $a = 2$ and $b = 3$

$$\frac{4 - a}{a - b}$$

4. Evaluate the algebraic expression for $a = 2$ and $b = -1$

$$3a - 2b^2 + (a - b)$$

5. Evaluate the algebraic expression for $q = -1.5$, $r = 1.5$, $s = -2$, and $t = 2$

$$q - 2r - \frac{1}{4}(t - 3s)$$

6. Evaluate the algebraic expression for $a = 3$, $b = -2$, $c = -0.5$, and $d = 1.5$

$$\frac{-6ab}{(c + d)^2}$$

7. Evaluate the algebraic expression for $a = -0.5$, $b = 3$, $c = -2$, and $d = 1.5$

$$b^2 - 2d + \frac{1}{3}(2a - c)$$

8. Evaluate the algebraic expression for $x = -3$ and $y = 4$

   $x^2 - 4y^2$

9. Evaluate the algebraic expression for $r = -1$ and $s = 2$

   $2r - 3s + 4$

10. Evaluate the algebraic expression for $a = 3$ and $b = -2$

   $4 - 2a + 5b$

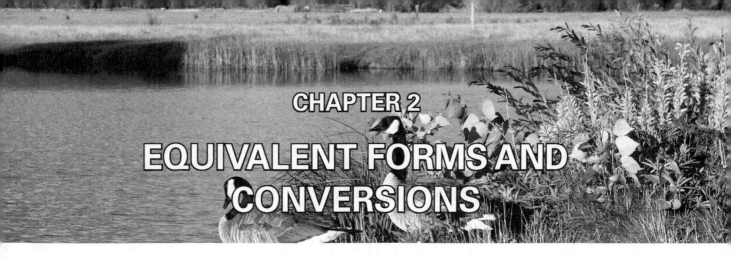

2.1 Multiplying and Dividing Fractions and Mixed Numbers

Multiplying fractions is not as complex as addition and subtraction of fractions because a common denominator is not needed. For both proper fractions (those with a numerator less than the denominator) and improper fractions (those with a numerator larger than the denominator), the process is quite straightforward. If you need to multiply or divide mixed fractions, one extra step is required.

Multiplying Proper or Improper Fractions:

Step 1: Multiply the numerators (top numbers) together. This is the new numerator.

Step 2: Multiply the denominators (bottom numbers) together. This is the new denominator.

EXAMPLE 1

Multiply:

$$\frac{4}{7} \cdot \frac{5}{12}$$

Step 1: Since $4 \cdot 5 = 20$, the new numerator is 20.

Step 2: Since $7 \cdot 12 = 84$, the new denominator is 84.

Step 3: The unreduced answer is 20/84. That is:

$$\frac{4}{7} \cdot \frac{5}{12} = \frac{20}{84}$$

This answer can be reduced by cancellation of shared prime factors. The TI83/84 calculator can also be used to reduce the final answer. When you put 20/84 into the calculator and push the buttons ENTER, MATH, ENTER, ENTER, the reduced fraction 5/21 is given.

Multiplying Mixed Numbers:

Before we can multiply mixed number fractions, we need to convert them into improper fractions. To do this, we need to recognize that the whole number in a mixed number can be seen as a fraction

15

with the same denominator as the fractional part of the mixed number. For example, consider the mixed number

$$2\frac{1}{3}$$

The picture below shows this number broken into thirds:

You can see that there are 7 thirds in $2\frac{1}{3}$. So

$$2\frac{1}{3} = \frac{7}{3}$$

An easy way to make this conversion is to multiply the whole number times the denominator and then add the value in the numerator to get the new numerator of the improper fraction.

$$2\frac{1}{3} = \frac{6+1}{3} = \frac{7}{3}$$
$$2\cdot3=6$$

EXAMPLE 2
Multiply:

$$5\frac{1}{3} \cdot 2\frac{1}{4}$$

Step 1: Because this problem involves the multiplication of mixed numbers, we must first change the mixed numbers into improper fractions.

$$5\frac{1}{3} = \frac{15+1}{3} = \frac{16}{3}$$
$$2\frac{1}{4} = \frac{8+1}{4} = \frac{9}{4}$$

Step 2: The problem can be rewritten as:

$$\frac{16}{3} \cdot \frac{9}{4}$$

Step 3: Since $16\cdot9 = 144$, the new numerator is 144.

Step 4: Since $3\cdot4 = 12$, the new denominator is 12.

Step 5: Completing the problem:

$$\frac{16}{3} \cdot \frac{9}{4} = \frac{144}{12} = 12$$

Dividing Fractions:

Dividing fractions is almost as easy as multiplying them. In fact, we change division problems into multiplication problems by "flipping over" the fraction we wish to divide by and then multiplying. This "flipped" fraction is called the reciprocal of the original fraction. Consider the following example.

EXAMPLE 3
Divide:

$$\frac{3}{4} \div \frac{2}{3}$$

This problem is read "3/4 divided by 2/3."

Step 1: Since we wish to divide by 2/3, we will multiply by the reciprocal, 3/2.

$$\frac{3}{4} \div \frac{2}{3} = \frac{3}{4} \cdot \frac{3}{2}$$

Step 2: Now we can proceed as we did in the first section and multiply the numerators and denominators to get the final answer.

$$\frac{3}{4} \div \frac{2}{3} = \frac{3}{4} \cdot \frac{3}{2} = \frac{9}{8}$$

EXAMPLE 4
Divide:

$$2\frac{1}{2} \div 4\frac{1}{3}$$

Step 1: Because this problem involves the multiplication of mixed numbers, we must first change the mixed numbers into improper fractions.

$$2\frac{1}{2} \div 4\frac{1}{3} = \frac{5}{2} \div \frac{13}{3}$$

Step 2: Since we wish to divide by 13/3, we will multiply by the reciprocal, 3/13.

$$2\frac{1}{2} \div 4\frac{1}{3} = \frac{5}{2} \div \frac{13}{3} = \frac{5}{2} \cdot \frac{3}{13}$$

Step 3: Now we can proceed as we did in the first section and multiply the numerators and denominators to get the final answer.

$$\frac{5}{2} \cdot \frac{3}{13} = \frac{15}{26}$$

2.1 PRACTICE PROBLEMS

1. Multiply the following fractions.

    a. $\dfrac{24}{66} \cdot \dfrac{2}{3} =$

    b. $\left(\dfrac{11}{59}\right)\left(\dfrac{6}{7}\right)$

    c. $\dfrac{25}{111} \cdot \dfrac{12}{23} \cdot \dfrac{2}{5} =$

    d. $2\dfrac{2}{3} \cdot \dfrac{5}{6} =$

    e. $3\dfrac{1}{15} \cdot 2\dfrac{2}{9} =$

2. Divide the following fractions.

    a. $\dfrac{5}{6} \div \dfrac{3}{4} =$

    b. $\dfrac{26}{35} \div \dfrac{3}{5} =$

    c. $\dfrac{5}{4} \div 2\dfrac{1}{3} =$

    d. $5\dfrac{1}{2} \div \dfrac{2}{7} =$

    e. $\dfrac{3}{5} \div 6 =$

2.2 Converting between Fractions, Decimals, and Percentages

It is often necessary in mathematics to convert between a fraction, decimal, and percentage. A review of these topics should make it possible for you to convert easily from one form to another.

Fractions:

The basic idea is that a fraction represents "parts of a whole." For example, ¼ represents one part out of a total of four parts:

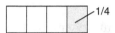

The bottom number in a fraction (the denominator) defines how many parts the whole is split into. The upper number in the fraction (the numerator) tells us the number of parts we are representing in the fraction. In the picture above, we split the whole rectangle into 4 equal parts but only 1 of those parts is shaded, so the fraction can be written as ¼.

Decimals:

To convert a fraction to a decimal, we must remember that a fraction also can be interpreted as the numerator divided by the denominator. Thus, ¼ = 1 ÷ 4 = 0.25. This is an example of a terminating decimal because it ends without rounding.

We can perform division to change any fraction into a decimal. But, not all fractions will result in terminating decimals. For example, 2/3 results in a repeating decimal: $0.6666\overline{6}$ where the bar over the last 6 indicates that the 6 would repeat indefinitely. We typically will use rounding rules to show this number with only a few decimal places. However, as soon as we round the value, we have chosen to give an approximate answer that is not exactly equal to the original fraction.

Converting a decimal to a fraction:

To perform this task, we must know the place value for the digit furthest to the right (the last place in the number). Thus, 0.569 has a 9 in the thousandths place so its fractional equivalent is 569/1000. Place values for several places are shown below:

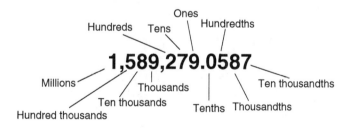

Percentage Conversions:

To convert a fraction to a percentage, we must first find the decimal equivalent. Then, we multiply by 100, which is the same as moving the decimal two places to the right, and add the percent sign. Therefore, 3/4 = 0.75 = 75%.

To convert from a percentage to a fraction, we must first get the decimal equivalent. We do this by dividing by 100, which is the same as moving the decimal two places to the left, and remove the percent sign. Then use the steps for converting a decimal to a fraction. Therefore, $67.2\% = 0.672 = \frac{672}{1,000}$, which is the unreduced fraction.

EXAMPLE 1

Convert 17/200 from a fraction to decimal to percentage:

Step 1: Convert to a decimal by dividing the fraction:

$$\frac{17}{200} = 17 \div 200 = 0.085$$

Step 2: Convert to a percentage by multiplying by 100:

$$0.085 = 0.085 \cdot 100\% = 8.5\%$$

EXAMPLE 2

Convert 38.4% from a percentage to a decimal and then a fraction:

$$38.4\% = 38.4 \div 100\% = 0.384$$

The four is in the thousandths place so the unreduced fraction is $0.384 = \frac{384}{1,000}$.

2.2 PRACTICE PROBLEMS

1. Fill in the following table by making all the necessary conversions.

Fraction (leave as an improper fraction if needed)	Decimal (round to four places where needed)	Percentage
3/5		
	0.589	
		56.2%
	1.36	
6/17		
		235%
		2.8%
	0.2987	

2. On one committee, four of the seven members are women.

    a. What fraction of the committee is made up of women?

    b. What decimal part of the committee is made up of women?

    c. What percentage of the committee is made up of women?

3. The newspaper report says that one-third of all voters in the state are Democrats.

    a. What decimal part of the voters are Democrats?

    b. What percent of the voters in the state are Democrats?

2.3 Performing Unit Conversions

The process of converting from one unit measurement to another is not complex and should not be dependent on any knowledge of the "size" of the units relative to one another. The process uses multiplication by "unit converters or unit fractions." This process is also referred to as unit analysis or dimensional analysis.

Unit Converter (or Unit Fractions):

Two unit converters can be written for any equivalency. For example, consider the equivalency between inches and feet:

$$12 \text{ in.} = 1 \text{ ft}$$

Because the numerators and denominators in each of these fractions are equal to each other the ratio is equivalent to 1:

$$\frac{12 \text{ in.}}{1 \text{ ft}} = \frac{1 \text{ ft}}{12 \text{ in.}} = 1$$

Both of the fractions shown above are unit converters. The choice of which unit converter to use is dependent on what units you desire to have in the result. You multiply by the unit converter that has the desired final units in the numerator.

Consider the following examples.

EXAMPLE 1

We have a roll of fencing that measures 1800 inches in length. How many feet of fencing do we have?

Step 1: First, identify that the answer should be in "feet." Therefore, we will multiply by the converter:

$$\frac{1 \text{ ft}}{12 \text{ in.}}$$

Step 2: Now multiply:

$$1800 \text{ in.} \cdot \frac{1 \text{ ft}}{12 \text{ in.}} = 150 \text{ ft}$$

Notice that the units "inches" cancel because they appear in both the numerator and the denominator.

EXAMPLE 2

We wish to convert 14 inches into centimeters.

Step 1: The equivalency between inches and centimeters is given below:

$$2.54 \text{ cm} = 1 \text{ in.}$$

Step 2: Use the information above to make the two unit fractions.

$$\frac{2.54 \text{ cm}}{1 \text{ in.}} = \frac{1 \text{ in.}}{2.54 \text{ cm}}$$

Step 3: Since we are changing inches into centimeters, we choose the unit fraction with centimeters in the numerator and multiply.

$$14 \text{ in.} \cdot \frac{2.54 \text{ cm}}{1 \text{ in.}} = 35.56 \text{ cm}$$

As in Example 1, the "inches" cancel and leave us with only centimeters.

EXAMPLE 3

How many kilograms are there in 2500 grams?

Step 1: The equivalency between grams and kilograms is given below:

$$1000 \text{ g} = 1 \text{ kg}$$

Step 2: Use the information above to make the two unit fractions.

$$\frac{1000 \text{ g}}{1 \text{ kg}} = \frac{1 \text{ kg}}{1000 \text{ g}}$$

Step 3: Since we are changing grams into kilograms, we choose the unit fraction with kilograms in the numerator and multiply.

$$2500 \text{ g} \cdot \frac{1 \text{ kg}}{1000 \text{ g}} = 2.5 \text{ kg}$$

EXAMPLE 4

How many square inches (in.^{2}) are there in 9.5 square feet (ft^2)?

Step 1: The equivalency between feet and inches is given below:

$$12 \text{ in.} = 1 \text{ ft}$$

Step 2: Notice the definition of "square inches" is **in.** $\cdot$ **in.** = **in.**^{2} The same is true for square feet.

Step 3: Because we wish to end up in square inches, we will choose the unit fraction that has inches in the numerator. However, since we are dealing with square inches, we need to multiply by the unit fraction twice.

$$9.5 \text{ ft}^2 \cdot \frac{12 \text{ in.}}{1 \text{ ft}} \cdot \frac{12 \text{ in.}}{1 \text{ ft}} = (9.5 \cdot 12 \cdot 12) \text{ in.}^2 = 1368 \text{ in.}^2$$

2.3 PRACTICE PROBLEMS

In Problems 1–10, determine the possible unit converters (unit fractions) for the equivalents listed.

1. 5280 ft = 1 mile

2. 1 km = 1000 m

3. 0.946 L = 1 qt

4. 100 cm = 1 m

5. 36 in. = 1 yd

6. 1 mile = 1.6 km

7. 1 in. = 2.54 cm

8. 1 ft = 30.48 cm

9. 1 yd = 0.9 m

10. 1 ml = 1000 L

Using the unit converters from Problems 1–10, change to the requested units. Show your work.

11. Ten kilometers (km) is how many miles?

12. How many liters (L) are in 2 qt?

13. How long is a 100-yard dash in inches?

14. 28 in. = _____ cm

15. 4 ft = _____ cm

16. How many meters (m) in 3 yards?

17.

Kilograms	Centigrams	Grams
0.00548		

18.

Milliliters	Liters	Kiloliters
		2.54

19. If you buy 100 square yards (yd^2) of carpet, how many square feet did you buy?

2.4 Converting between Scientific Notation and Standard Notation

Scientific notation was developed to allow us to write and perform operations with very large or very small numbers more easily.

A number is in scientific notation when it is broken up as the product of two parts: a number, a, which is greater than or equal to 1 and less than 10, multiplied by 10 raised to an integer power, n. Symbolically, this format is $a \cdot 10^n$.

Converting from standard notation to scientific notation:

There are three steps to consider when converting a number from standard notation to scientific notation:

Step 1: Place the decimal point such that there is one nonzero digit to the left of the decimal point.

Step 2: Count the number of decimal places that the decimal has "moved" from the original number. This will determine the exponent of the 10.

Step 3: If the original number is less than 1, the exponent is negative; if the original number is greater than 1, the exponent is positive.

EXAMPLE 1

Convert 5268 from standard notation to scientific notation:

$$5268 = 5268.0 = 5.268 \times 10^3$$

To get one nonzero digit to the left of the decimal place, you will need to move the decimal point (which is understood to be at the far right of the number to start out) three places to the left. Therefore, the exponent is a 3. This also means that the number a in this example is 5.268. Since the original number was greater than 1, the exponent in scientific notation is positive. Thus, the answer:

$$5.268 \cdot 10^3$$

EXAMPLE 2

Convert 0.00689 to scientific notation

$$0.00689 = 006.89 = 6.89 \times 10^{-3}$$

Note that the exponent in this example is negative because the original number in standard notation was less than 1.

Converting from scientific notation to standard notation:
There are two steps to consider when converting from scientific notation to standard notation:

> **Step 1:** If the exponent is positive, move the decimal point to the right. The exponent tells you how many places to move the decimal point.

> **Step 2:** Or, if the exponent is negative, move the decimal point to the left. Again, the exponent tells you how many places to move the decimal point.

EXAMPLE 3

Convert from scientific notation to standard form:

$$7.21 \cdot 10^3 = 7210$$

Because the exponent is positive, move the decimal point to the right three places. The resulting number, 7210, is greater than 1 as expected.

Most calculators, including the TI-83/84, will automatically express very small and very large numbers in scientific notation. In order to understand the display, consider the following example.

If the calculator display shows

$$2.679E2$$

Then the number following the "E" is the exponent of 10. The full translation is:

$$2.679E2 = 2.679 \cdot 10^2$$

EXAMPLE 4

Write the calculator display $5.697E - 4$ in scientific and standard notations.

$$5.697 \cdot 10^{-4} = 0.0005697$$

Because the exponent is negative, move the decimal point to the left four places. You must put zeroes in as "place holders" as needed. The resulting answer of 0.0005697 is less than 1 as expected.

2.4 PRACTICE PROBLEMS

1. Fill in the following table by making all the necessary conversions.

Standard Notation	Scientific Notation
560,000	
	$2.8946 \cdot 10^2$
	$6.5 \cdot 10^{-5}$
123.89	
	$2.7 \cdot 10^{-4}$
0.1589	
	$5.97 \cdot 10^8$
0.00057	
0.00000079	

2. Put the following calculator displays into scientific notation and then standard notation.

Calculator Display	Scientific Notation	Standard Notation
3.265E1		
2.79$E-5$		
4.587E6		
5.222$E-8$		

3. The radius of the Sun is 70 billion centimeters. Write this number in scientific notation.

4. The length of a flu virus is 0.00000013 m. Write this number in scientific notation.

UNIT II
PRINCIPLES OF RATIONAL NUMBERS, EXPONENTS, AND PERSONAL FINANCE

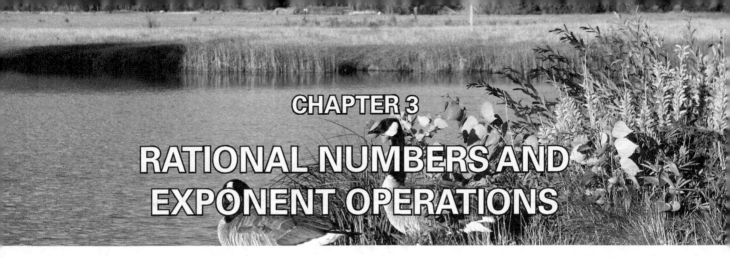

CHAPTER 3
RATIONAL NUMBERS AND EXPONENT OPERATIONS

3.1 Using the Product Rule for Exponents

Exponent Review:

An exponent operation looks like this:

$$\text{Exponent} \Downarrow$$
$$3^4 = 3 \cdot 3 \cdot 3 \cdot 3 = 81$$
$$\Uparrow \text{ Base}$$

The exponent is a number that indicates how many times the base number is multiplied by itself. In the example above, the base number 3 was multiplied by itself four times. Note that any number has an understood exponent of 1, which we never write. That is, $3^1 = 3$ or $x^1 = x$.

The base can be a variable, so $y^3 = y \cdot y \cdot y$.

Product Rule:

We can simplify exponent multiplications when the bases are the same by using the product rule for exponents. This rule states:

If x is a nonzero real number and m and n are integers, then:

$$x^m \cdot x^n = x^{m+n}$$

That is, if you are multiplying powers with the same base, you simply add the exponents and keep the base. This rule is known as the product rule because the answer to multiplications are called products.

EXAMPLE 1
Simplify:

$$x^8 x^2$$

Because both bases are x, we can use the product rule to simplify.

$$x^8 x^2 = x^{8+2} = x^{10}$$

EXAMPLE 2
Simplify:

$$x^{-3}x^3$$

Step 1: Because the product rule is defined for all integer exponents, the exponents can be negative. Therefore:

$$x^{-3}x^3 = x^{-3+3} = x^0$$

Step 2: The value of any number raised to the power 0 equals 1. Therefore:

$$x^{-3}x^3 = x^{-3+3} = x^0 = 1$$

EXAMPLE 3
Simplify:

$$x^3x^2y^2$$

We can use the product rule to combine the x factors, but we can't include the y factor because the base is different.

$$x^3x^2y^2 = x^{3+2}y^2 = x^5y^2$$

EXAMPLE 4
Simplify:

$$(3x^2y^3)(5x^2y^2)$$

Step 1: Because multiplication is commutative, we can group the factors with the same bases:

$$(3x^2y^3)(5x^2y^2) = (3 \cdot 5)(x^2x^2)(y^3y^2)$$

Step 2: We can now use the product rule to simplify.

$$(3x^2y^3)(5x^2y^2) = (3 \cdot 5)(x^2x^2)(y^3y^2) = 15x^4y^5$$

3.1 PRACTICE PROBLEMS

1. $x^3x^5 =$

2. $y^{-2}y^6 =$

3. $x \cdot x^7 =$

4. $y^3y^2y^4 =$

5. $a^3a^{-1} =$

6. $a^2a^4a^3 =$

7. $b^{-3}b =$

8. $x^{-2}x^2 =$

9. $(4a^3b^4)(2a^2b^3) =$

10. $(6x^5y)(-2x^{-1}y^2) =$

3.2 Using the Quotient Rule for Exponents

We frequently need to evaluate expressions where exponents are in fractions. Consider the example below:

$$\frac{x^4}{x^3} = \frac{\cancel{x} \cdot \cancel{x} \cdot \cancel{x} \cdot x}{\cancel{x} \cdot \cancel{x} \cdot \cancel{x}} = x^1 = x$$

By showing the explicit meaning of the exponents and using cancellation, we conclude that the answer is simply x.

The quotient rule is simply a shorthand way of doing the expansion and cancellation shown above. It takes into account the fact that fractions indicate division of the numerator by the denominator.

Quotient Rule:

If x is a nonzero real number and m and n are integers, then:

$$\frac{x^m}{x^n} = x^{m-n}$$

Note that the powers must have the same base and that the exponent in the denominator is subtracted from the exponent in the numerator.

EXAMPLE 1
Simplify.

$$\frac{3^5}{3^2}$$

Because there is the shared base number, 3, we can use the quotient rule to simplify.

$$\frac{3^5}{3^2} = 3^{5-2} = 3^3 = 27$$

EXAMPLE 2
Simplify.

$$\frac{x^4 z^2}{x^3}$$

We can simplify the powers using the quotient rule:

$$\frac{x^4 z^2}{x^3} = x^{(4-3)} z^2 = x^1 z^2 = xz^2$$

EXAMPLE 3
Simplify.

$$\frac{15x^5r^3}{5r^4x^3}$$

Step 1: We can use the associative and commutative rules to regroup the divisions.

$$\frac{15x^5r^3}{5r^4x^3}=\frac{15}{5}\cdot\frac{x^5}{x^3}\cdot\frac{r^3}{r^4}$$

Step 2: Now we can use the product rule to simplify.

$$\frac{15x^5r^3}{5r^4x^3}=\frac{15}{5}\cdot\frac{x^5}{x^3}\cdot\frac{r^3}{r^4}=3x^{5-3}r^{3-4}=3x^2r^{-1}$$

However, we usually don't like to write an expression with negative exponents. So, we note that the following is true:

If x is a nonzero number and n is an integer, then:

$$x^{-n}=\frac{1}{x^n}$$

Step 3: We can use the rule above to rewrite the answer.

$$3x^2r^{-1}=\frac{3x^2}{r^1}=\frac{3x^2}{r}$$

3.2 PRACTICE PROBLEMS

1. $\dfrac{y^6}{y^3} =$

2. $\dfrac{x^5}{x^2} =$

3. $\dfrac{x^4 y^2}{x} =$

4. $\dfrac{14 x^3 y^4}{7 x^2 y^5} =$

5. $\dfrac{25 a^5 b^4}{5 a^2 b^2} =$

6. $\dfrac{20 x^6 y^6}{4 x^2 y^3} =$

7. $4 x^2 y^{-2} =$

8. $5 x^{-3} y^2 =$

3.3 Multiplying and Dividing with Scientific Notation

Recall that scientific notation is simply a shorthand way to show very large or small numbers. The format of scientific notation is the following:

$$a \cdot 10^n$$

where a is a real number between 1 and 10 and n is an integer.

When we have to do multiplications and divisions with numbers in scientific notation, we need to use the product and quotient rules studied in the previous sections.

EXAMPLE 1
Simplify.

$$(2.3 \cdot 10^3)(5.6 \cdot 10^6)$$

Step 1: Regroup.

$$(2.3 \cdot 10^3)(5.6 \cdot 10^6) = (2.3 \cdot 5.6)(10^3 \cdot 10^6)$$

Step 2: Multiply and use the product rule.

$$(2.3 \cdot 5.6)(10^3 \cdot 10^6) = 12.88 \cdot 10^{3+6} = 12.88 \cdot 10^9$$

Step 3: We are not finished because the 12.88 is not between 1 and 10 as required by the rules of scientific notation. To change 12.88 into scientific notation, move the decimal place over one place to the left: $\mathbf{1.288 \cdot 10^1}$.

Now substitute:

$$12.88 \cdot 10^9 = 1.288 \cdot 10^1 \cdot 10^9 = 1.288 \cdot 10^{10}$$

EXAMPLE 2
Simplify.

$$\frac{5.3 \cdot 10^{-3}}{4 \cdot 10^{-6}}$$

Step 1: Regroup.

$$\frac{5.3}{4} \cdot \frac{10^{-3}}{10^{-6}}$$

Step 2: Divide and use the quotient rule.

$$\frac{5.3}{4} \cdot \frac{10^{-3}}{10^{-6}} = 1.325 \cdot 10^{-3-(-6)} = 1.325 \cdot 10^3$$

EXAMPLE 3

Simplify.

$$\frac{(5 \cdot 10^2)(3.2 \cdot 10^2)}{4 \cdot 10^4}$$

Step 1: Regroup.

$$\frac{(5 \cdot 10^2)(3.2 \cdot 10^2)}{4 \cdot 10^4} = \frac{5 \cdot 3.2}{4} \cdot \frac{10^2 \cdot 10^2}{10^4}$$

Step 2: Multiply and divide and use the product and quotient rules.

$$\frac{5 \cdot 3.2}{4} \cdot \frac{10^2 \cdot 10^2}{10^4} = 4 \cdot \frac{10^{2+2}}{10^4} = 4 \cdot \frac{10^4}{10^4} = 4 \cdot 10^{4-4} = 4 \cdot 10^0 = 4 \cdot 1 = 4$$

3.3 PRACTICE PROBLEMS

1. $\left(4\cdot10^{3}\right)\left(2\cdot10^{5}\right)=$

2. $\left(5\cdot10^{-4}\right)\left(2.3\cdot10^{-1}\right)=$

3. $\left(1.5\cdot10^{-6}\right)\left(3.0\cdot10^{4}\right)=$

4. $\left(2.1\cdot10^{6}\right)\left(2.1\cdot10^{-4}\right)=$

5. $\dfrac{4\cdot10^{3}}{2\cdot10^{5}}=$

6. $\dfrac{5\cdot10^{-4}}{2.3\cdot10^{-1}}=$

7. $\dfrac{1.5\cdot10^{-6}}{3.0\cdot10^{4}}=$

8. $\dfrac{2.1\cdot10^{6}}{7\cdot10^{-4}}=$

3.4 Rounding and Place Value

Place Value:

In order to round correctly, you need to have knowledge of "place value" for each of the digits in the number. The following chart is a "map" of place value with special emphasis on the places to the right of the decimal place.

Place Value Chart													
Millions	Hundred thousands	Ten thousands	Thousands	Hundreds	Tens	Ones	• Decimal point	Tenths	Hundredths	Thousandths	Ten thousandths	Hundred thousandths

Notice that there is a "ones" place to the left of the decimal, but there is no "oneth" place to the right!

Process of Rounding Numbers:

Step 1: Decide which place value you must round to.

Step 2: Look at the number in the place to the right.

- If the value in this place is less than five (0 to 4), there is no change to the place value.

- If the value in this next place is greater than or equal to five (5 to 9), there is a change of +1 to the place value.

EXAMPLE 1

Round 0.5878 to the hundredths place (two decimal places):

Step 1: The hundredths place is the second place to the right of the decimal. There is an 8 is this place.

Step 2: Look at the next place to the right; there is a 7 here. Since the 7 is in the 5 to 9 range, we round the 8, which was in the hundredths place to a 9.

Answer: The rounded number is 0.59.

Decimals to Fractions:

We may be asked to express a decimal as a fraction. To do this, we also have to understand place value because the denominator of the fraction is determined by the place value of the number furthest right of the decimal. In probabilities, all of the numbers will be between 0 and 1. Therefore, we don't need to worry about numbers to the left of the decimal. We may also need to reduce the fraction but this can easily be done using the calculator.

To write a decimal as a fraction do the following:

> **Step 1:** Identify the place value of the digit furthest to the right.

> **Step 2:** Write this number as the denominator (the bottom number) of a fraction.

> **Step 3:** The numerator of the fraction is simply the digits to the right of the decimal place.

EXAMPLE 2
Change 0.126 to a fraction:

> **Step 1:** The number furthest to the right is the 6, which is in the thousandths place. So the denominator of the fractional form is 1000.

> **Step 2:** The numerator of the fraction is simply the digits to the right of the decimal place: 126.

> **Answer:** The full fraction would be $\frac{126}{1,000}$, which can be reduced to $\frac{63}{500}$.

> Notice that the fraction form has NO decimal point.

3.4 PRACTICE PROBLEMS

Round to the tenths place for Problems 1 and 2.

1. 32.6543

2. 182.5463

Round to the hundredths place for Problems 3 and 4.

3. 15.81234

4. 6.19894

Change decimal number to a simplified fraction for Problems 5–7.

5. 0.375

6. 2.15

7. 0.0125

8. Round 3.1428137853 to 7 decimal places.

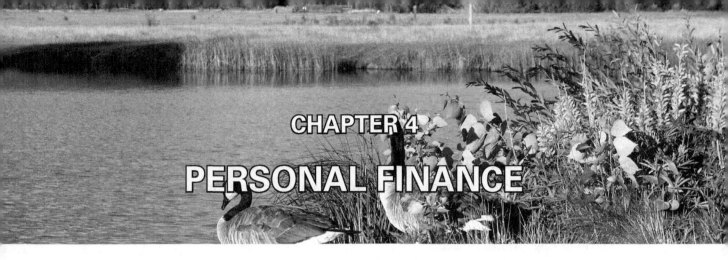

4.1 Evaluating Financial Formulas Involving Powers

For many students, using financial formulas is a significant challenge. Students have problems identifying the numerical values associated with the symbols in these equations. They also have issues following the order of operations correctly in calculations involving these formulas.

Table of Symbols Commonly Used in Financial Formulas:

Symbol	Meaning
A	Accumulated balance after Y years
P	Starting principal amount
APR	Annual percentage rate of interest (as a decimal)
Y	Number of years
n	The number of compounding periods per year or the number of payments per year
PMT	Payment amount

EXAMPLE 1

If you invest $1000 at 4% compounded monthly for 5 years, what will your balance be in the account at the end of the 5 years? Identify the variables first.

The principal amount is the amount you start with. Therefore, $P = 1000$.

The annual percentage rate, APR, is 4%. However, this value must be written as a decimal number. Therefore, $APR = 0.04$.

The bank compounds the amount monthly. Since there are 12 months in a year, $n = 12$.

The investment lasts 5 years. Therefore, $Y = 5$.

We are solving for the amount in the account at the end of the 5 years, which is A. This means that we should look for a formula that solves for A using the other variables.

Order of Operations in Evaluating Financial Formulas:

Let's look at the completion of Example 1, concentrating on the order of operations.

Step 1: First, we need to find the correct formula.

The following formula is needed:

$$A = P\left(1 + \frac{APR}{n}\right)^{nY}$$

Step 2: We have already identified all of the variables in Example 1. Now substitute into the formula above.

$$A = 1000\left(1 + \frac{0.04}{12}\right)^{12 \cdot 5}$$

Step 3: Recalling the order of operations, the operations inside the parenthesis must be done first. Be sure to do the division before the addition!

$$A = 1000\left(1 + \frac{0.04}{12}\right)^{12 \cdot 5} = 1000(1.003333)^{12 \cdot 5}$$

Step 4: Notice that there is an implied parenthesis around the exponent, so we multiply those numbers together to see the value of the exponent.

$$A = 1000(1.003333)^{12 \cdot 5} = 1000(1.003333)^{60}$$

Step 5: Now evaluate the exponent operation.

$$A = 1000 \cdot 1.22097$$

Step 6: Finally, do the multiplication.

$$A = 1000 \cdot 1.22097 = 1220.97$$

That is, we would have \$1220.97 in the account at the end of 5 years.

EXAMPLE 2

If I deposit \$4000 into an account that compounds continuously and leave it there at 5% for 10 years, how much money will I have?

Step 1: Identify the formula needed. Since we are solving for the balance in the account, A, with continuous compounding, we need the formula below.

$$A = Pe^{(APR \cdot Y)}$$

Step 2: Identify the variables:

$$P = 4000$$
$$APR = 0.05$$
$$Y = 10$$

Step 3: Substitute into the equation.

$$A = 4000e^{(0.05 \cdot 10)}$$

Step 4: First, we must do the multiplication in the parentheses to get the exponent value.

$$A = 4000e^{(0.05 \cdot 10)} = 4000e^{0.5}$$

Step 5: Next, we evaluate the exponent operation.

$$A = 4000e^{0.5} = 4000 \cdot 1.64872$$

Step 6: Finally, do the multiplication.

$$A = 4000 \cdot 1.64872 = 6594.88$$

That is, we would have $6594.88 in the account at the end of 10 years.

4.1 PRACTICE PROBLEMS

For Problems 1–4, use the formula: $A = P\left(1 + \dfrac{APR}{n}\right)^{nY}$

1. $ 12,000 is invested for 5 years at an APR of 2% compounded weekly.
   a) Identify the variables values.
      - P (starting principal amount) =
      - APR (annual percentage rate of interest as a decimal) =
      - n (number of compounding periods per year) =
      - Y (number of years) =

   b) Consider the order of operations in applying this formula. Circle the step you would do first.
      - Do operations in the parenthesis.
      - Evaluate the exponent.
      - Do the multiplication.

   c) Circle the step you would do second.
      - Do operations in the parenthesis.
      - Evaluate the exponent.
      - Do the multiplication.

   d) Circle the step you would do last.
      - Do operations in the parenthesis.
      - Evaluate the exponent.
      - Do the multiplication.

   e) Complete the problem by calculating the accumulated balance in the account. Show all your work. Do not round until the final answer that should be rounded to the nearest cent.

      $A =$

2. Fifteen hundred dollars is invested for 12 years at an APR of 1.2%, compounded quarterly.
   a) Identify the variables values.
      - P (starting principal amount) =
      - APR (annual percentage rate of interest as a decimal) =
      - n (number of compounding periods per year) =
      - Y (number of years) =

   b) Consider the order of operations in applying the formula. What would you do first?

   c) What would you do next?

   d) What would you do last?

   e) Complete the problem by calculating the accumulated balance in the account. Show all your work. Do not round until the final answer that should be rounded to the nearest cent.

      $A =$

3. Sally invested $12,500 for 3.5 years at an *APR* of 2% compounded monthly.
    a) Identify the variables values.
       - *P* (starting principal amount) =
       - *APR* (annual percentage rate of interest as a decimal) =
       - *n* (number of compounding periods per year) =
       - *Y* (number of years) =

    b) Consider the order of operations in applying the formula. What would you do first?

    c) What would you do next?

    d) What would you do last?

    e) Complete the problem by calculating the accumulated balance in the account. Show all your work. Do not round until the final answer that should be rounded to the nearest cent.

       $A =$

4. Fifty-five hundred dollars is invested for 6 years at an *APR* of 10% compounded daily.
    a) Identify the variables values.
       - *P* (starting principal amount) =
       - *APR* (annual percentage rate of interest as a decimal) =
       - *n* (number of compounding periods per year) =
       - *Y* (number of years) =

    b) Consider the order of operations in applying the formula. What would you do first?

    c) What would you do next?

    d) What would you do last?

    e) Complete the problem by calculating the accumulated balance in the account. Show all your work. Do not round until the final answer that should be rounded to the nearest cent.

       $A =$

For Problems 5–8, use the formula: $A = Pe^{(APR \cdot Y)}$

5. You deposit $5000 in an account paying 2% interest for 5 years, compounded continuously.
    a) Identify the variables values.
       - *P* (starting principal amount) =
       - *APR* (annual percentage rate of interest as a decimal) =
       - *Y* (number of years) =

    b) Consider the order of operations in applying this formula. Circle the step you would do first.
       - Do operations in the parenthesis.
       - Evaluate the exponent.
       - Do the multiplication.

c) Circle the step you would do next.
- Do operations in the parenthesis.
- Evaluate the exponent.
- Do the multiplication.

d) Circle the step you would do last.
- Do operations in the parenthesis.
- Evaluate the exponent.
- Do the multiplication.

e) Complete the problem by calculating the accumulated balance in the account. Show all your work. Do not round until the final answer that should be rounded to the nearest cent.

$A =$

6. You deposit $6500 in an account paying 3.5% interest for 4 years, compounded continuously.
a) Identify the variables values.
- P (starting principal amount) =
- APR (annual percentage rate of interest as a decimal) =
- Y (number of years) =

b) Consider the order of operations in applying the formula. What would you do first?

c) What would you do next?

d) What would you do last?

e) Complete the problem by calculating the accumulated balance in the account. Show all your work. Do not round until the final answer that should be rounded to the nearest cent.

$A =$

7. You deposit $5500 in an account paying 1.5% interest for 3 years, compounded continuously.
a) Identify the variables values.
- P (starting principal amount) =
- APR (annual percentage rate of interest as a decimal) =
- Y (number of years) =

b) Consider the order of operations in applying the formula. What would you do first?

c) What would you do next?

d) What would you do last?

e) Complete the problem by calculating the accumulated balance in the account. Show all your work. Do not round until the final answer that should be rounded to the nearest cent.

$A =$

8. You deposit $10,000 in an account paying 2% interest for 8.5 years, compounded continuously.
    a) Identify the variables values.
        - P (starting principal amount) =
        - APR (annual percentage rate of interest as a decimal) =
        - Y (number of years) =

    b) Consider the order of operations in applying the formula. What would you do first?

    c) What would you do next?

    d) What would you do last?

    e) Complete the problem by calculating the accumulated balance in the account. Show all your work. Do not round until the final answer that should be rounded to the nearest cent.

        $A =$

4.2 Evaluating Financial Formulas Containing Complex Numerators and Denominators

An additional challenge for students occurs when the financial formulas contain complex numerators, denominators, or both. In many of these cases, there will be parentheses inside parentheses. We called these "nested" parentheses; and they must be performed from the innermost parentheses out. Consider the examples below.

EXAMPLE 1

If we put $200 into a savings account each month which has a 4% interest rate, compounded quarterly, how much would we have in the account after 10 years?

The formula that I need to use is given below:

$$A = PMT \cdot \frac{\left[\left(1 + \dfrac{APR}{n}\right)^{(nY)} - 1\right]}{\left(\dfrac{APR}{n}\right)}$$

Step 1: Identify the variables:

$$PMT = 200$$
$$APR = 0.04$$
$$Y = 10$$
$$n = 4$$

Step 2: Substitute into the equation.

$$A = PMT \cdot \frac{\left[\left(1 + \dfrac{APR}{n}\right)^{(nY)} - 1\right]}{\left(\dfrac{APR}{n}\right)} = 200 \cdot \frac{\left[\left(1 + \dfrac{0.04}{4}\right)^{(4 \cdot 10)} - 1\right]}{\left(\dfrac{0.04}{4}\right)}$$

Step 3: Begin by doing the operations within the innermost parenthesis. Be sure to the division before the addition.

$$A = 200 \cdot \frac{\left[\left(1 + \dfrac{0.04}{4}\right)^{(4 \cdot 10)} - 1\right]}{\left(\dfrac{0.04}{4}\right)} = 200 \cdot \frac{\left[(1.01)^{40} - 1\right]}{0.01}$$

Note that we have done the operations within all of the inner parentheses.

Step 4: Now perform the exponent operation in the outer parenthesis or brackets.

$$A = 200 \cdot \frac{\left[(1.01)^{40} - 1\right]}{0.01} = 200 \cdot \frac{[1.4888637 - 1]}{0.01}$$

Step 5: Do the operations within this bracket.

$$A = 200 \cdot \frac{[1.4888637 - 1]}{0.01} = 200 \cdot \frac{0.4888637}{0.01}$$

Step 6: Finally, perform the multiplication and division from left to right.

$$A = 200 \cdot \frac{0.4888637}{0.01} = 9777.27$$

You would have $9,777.27 in the account after 10 years.

EXAMPLE 2

Pam wishes to buy a $16,000 car at 3% over 5 years. Pam will be making monthly payments. What will the amount of the payment be?

The formula that we need to use is given below:

$$PMT = \frac{P \cdot \left(\dfrac{APR}{n} \right)}{\left[1 - \left(1 + \dfrac{APR}{n} \right)^{(-nY)} \right]}$$

Step 1: Identify the variables:

$$P = 16000$$
$$APR = 0.03$$

$$n = 12 \text{ (number of payments made per year)}$$

$$Y = 5$$

Step 2: Substitute into the equation.

$$PMT = \frac{P \cdot \left(\dfrac{APR}{n} \right)}{\left[1 - \left(1 + \dfrac{APR}{n} \right)^{(-nY)} \right]} = \frac{16000 \cdot \left(\dfrac{0.03}{12} \right)}{\left[1 - \left(1 + \dfrac{0.03}{12} \right)^{(-12 \cdot 5)} \right]}$$

Step 3: Begin by doing the operations within the innermost parentheses. Be sure to the division before the addition.

$$PMT = \frac{16000 \cdot \left(\dfrac{0.03}{12} \right)}{\left[1 - \left(1 + \dfrac{0.03}{12} \right)^{(-12.5)} \right]} = \frac{16000 \cdot 0.0025}{\left[1 - 1.0025^{(-60)} \right]}$$

Step 4: Now perform the exponent operation.

$$PMT = \frac{16000 \cdot 0.0025}{[1 - 0.860869]}$$

Step 5: Do the operations within the bracket.

$$PMT = \frac{16000 \cdot 0.0025}{[1 - 0.860869]} = \frac{16000 \cdot 0.0025}{0.139131}$$

Step 6: Finally, perform the multiplication and division from left to right.

$$PMT = \frac{16000 \cdot 0.0025}{0.139131} = 287.50$$

Pam will need to pay $287.50 per month for 5 years.

4.2 PRACTICE PROBLEMS

For Problems 1–4, use the formula: $A = PMT \cdot \dfrac{\left[\left(1 + \dfrac{APR}{n}\right)^{nY} - 1\right]}{\left(\dfrac{APR}{n}\right)}$

1. Determine the amount that would be accumulated if we deposit $350 into an account each month at an *APR* of 5%, compounded quarterly, for 20 years.
   a) Identify the variable values.
      - *PMT* (payment amount) =
      - *APR* (annual percentage rate of interest as a decimal) =
      - *n* (number of payments per year) =
      - *Y* (number of years) =

   b) Consider the order of operations in applying this formula. Circle the step you would do first.
      - Perform operations in the outer parenthesis.
      - Perform operations in the inner parenthesis.
      - Evaluate the exponent.
      - Do the division.

   c) Circle the step you would do second.
      - Perform operations in the outer parenthesis.
      - Do the multiplication.
      - Evaluate the exponent.
      - Do the division.

   d) Circle the step you would do next.
      - Perform operations in the outer parenthesis.
      - Do the multiplication.
      - Evaluate the exponent.
      - Do the division.

   e) Circle the next step.
      - Divide by $\left(\dfrac{APR}{n}\right)$.
      - Multiply by $\left(\dfrac{APR}{n}\right)$.
      - Evaluate the exponent.

   f) Circle the last step.
      - Multiply by the *PMT*.
      - Divide by the *PMT*.

   g) Complete the problem by calculating the accumulated balance in the account. Show all your work. Do not round until the final answer that should be rounded to the nearest cent.

      $A =$

2. Determine the amount accumulated when you deposit $1500 into an account bimonthly at an *APR* of 4.5% for 10 years.
   a) Identify the variable values.
      - *PMT* (payment amount) =
      - *APR* (annual percentage rate of interest as a decimal) =
      - *n* (number of payments per year) =
      - *Y* (number of years) =

   b) Complete the problem by calculating the accumulated balance in the account. Show all your work. Do not round until the final answer that should be rounded to the nearest cent.

      $A =$

3. Determine the amount accumulated when you deposit $350 into an account each month with an *APR* of 1.5% for 25 years.
   a) Identify the variable values.
      - *PMT* (payment amount) =
      - *APR* (annual percentage rate of interest as a decimal) =
      - *n* (number of payments per year) =
      - *Y* (number of years) =

   b) Complete the problem by calculating the accumulated balance in the account. Show all your work. Do not round until the final answer that should be rounded to the nearest cent.

      $A =$

4. Determine the amount accumulated when you deposit $250 into an account each month with an *APR* of 4.5% for 30 years.
   a) Identify the variable values.
      - *PMT* (payment amount) =
      - *APR* (annual percentage rate of interest as a decimal) =
      - *n* (number of payments per year) =
      - *Y* (number of years) =

   b) Complete the problem by calculating the accumulated balance in the account. Show all your work. Do not round until the final answer that should be rounded to the nearest cent.

      $A =$

For Problems 5–6, use the formula: $PMT = \dfrac{A \cdot \left(\dfrac{APR}{n} \right)}{\left[\left(1 + \dfrac{APR}{n} \right)^{(nY)} - 1 \right]}$

5. How much should you deposit monthly with an *APR* of 2.5% to accumulate $65,000 in 25 years?
   a) Identify the variable values.
      - *A* (accumulated amount desired) =
      - *APR* (annual percentage rate of interest as a decimal) =
      - *n* (number of payments per year) =
      - *Y* (number of years) =

b) Consider the order of operations in applying this formula. Circle the step you would do first.
- Perform operations in the outer parenthesis.
- Perform operations in the inner parenthesis.
- Evaluate the exponent.
- Do the division.

c) Circle the step you would do second.
- Perform operations in the outer parenthesis.
- Do the multiplication.
- Evaluate the exponent.
- Do the division.

d) Circle the step you would do next.
- Perform operations in the outer parenthesis.
- Do the multiplication.
- Evaluate the exponent.
- Do the division.

e) Circle the next step.
- Divide by $\left(\dfrac{APR}{n} \right)$.
- Multiply $A \cdot \left(\dfrac{APR}{n} \right)$.
- Evaluate the exponent.

f) Circle the last step.
- Multiply by the denominator.
- Divide by the denominator.

g) Complete the problem by calculating the needed payment into the account. Show all your work. Do not round until the final answer that should be rounded to the nearest cent.

$PMT =$

6. How much should you deposit monthly with an APR of 3.45% to accumulate $100,000 in 10 years?
   a) Identify the variable values.
   - A (accumulated amount desired) =
   - APR (annual percentage rate of interest as a decimal) =
   - n (number of payments per year) =
   - Y (number of years) =

   b) Complete the problem by calculating the needed payment into the account. Show all your work. Do not round until the final answer that should be rounded to the nearest cent.

   $PMT =$

4.3 Solving Real-World Problems

The complexity of financial problems is compounded by the various forms that these problems may take. Many students find it difficult to parse out what the scenario means and identify the variables in the formulas. For example, the formulas use *PMT* to indicate a regular payment, but this may be a payment to yourself in the form of a deposit to a savings account or it may be a payment on a loan to a creditor. For students, the difference is substantial. Consider the following examples.

EXAMPLE 1

Problems that compare:
Because we live in a society with financial choices, we often need to compare options. This type of problem confuses because it is really *two* problems disguised as one.

Sue invests $1000 at her family's bank at 4% compounded monthly for 5 years. Ann decides to put $1000 in a different bank offering 3.75% compounded continuously for 5 years. Which student made the better decision?

The comparison must be based on which student ends up with the most money. Therefore, we need to solve for the amount accumulated in the account, A. The formulas needed are similar, but not exactly the same.

Step 1: Identify the variable for each case.

Variable	Sue's case	Ann's case
P, principal (amount invested)	$1000	$1000
APR (interest rate as decimal)	0.04	0.0375
n (number of compound periods per year)	12	continuous
Y (years of investment)	5	5

Step 2: In Sue's case, the formula needed is given below:

$$A = P \cdot \left(1 + \frac{APR}{n}\right)^{(nY)}$$

Step 3: Substituting into the formula and solving for Sue's case:

$$A = P \cdot \left(1 + \frac{APR}{n}\right)^{(nY)} = 1000 \cdot \left(1 + \frac{0.04}{12}\right)^{(12 \cdot 5)} = 1000(1.00333)^{60} = \$1220.70$$

Step 4: Because Ann's case involves continuous compounding, a different formula is used to calculate the amount accumulated, A. The formula needed is given below:

$$A = P \cdot e^{(APR \cdot Y)}$$

Step 5: Using the values from the table above for Ann's case, substitute and solve.

$$A = P \cdot e^{(APR \cdot Y)} = 1000 \cdot e^{(0.0375 \cdot 5)} = \$1206.23$$

It turns out that Sue made the better deal; but we needed to do the math to determine that.

EXAMPLE 2

Problems that confuse:

Traditionally, we think of payments (*PMT*) as an amount of money we pay someone else. However, it can be an amount we regularly pay ourselves in investment instruments such as savings accounts. Also, we usually think of amount accumulated (*A*) as an unknown future amount. However, it can be a specific amount that we will need in the future. In that case, *A* is known. Consider the following problem.

I need to have $150,000 available in 15 years for my child's college education. Assuming a 5% APR, how much do I need to deposit each month to meet this goal?

Step 1: We are trying to solve for the regular payment we make to the investment account. We can identify the variables as follows:

$$A = 150,000$$
$$Y = 15$$
$$APR = 0.05$$
$$n = 12$$

Notice that *n* here is not a compounding period, but instead it is the number of times per year that we make a payment into the account.

Step 2: Identify the formula needed. The formula must solve for *PMT* using the other identified variables. The formula needed is given below.

$$PMT = A \cdot \frac{\left(\dfrac{APR}{n}\right)}{\left[\left(1 + \dfrac{APR}{n}\right)^{(nY)} - 1\right]}$$

Step 3: Substituting and solving: (intermediate answers have been rounded to six decimal places).

$$PMT = 150,000 \cdot \frac{\left(\dfrac{0.05}{12}\right)}{\left[\left(1 + \dfrac{0.05}{12}\right)^{(12 \cdot 15)} - 1\right]} = 150,000 \cdot \frac{0.004167}{1.113830}$$

$$= \$561.17$$

So, I need to make a payment of $561.17 each month to reach my goal.

4.3 PRACTICE PROBLEMS

1. How much money would you have in an account if you deposit $2500 compounded quarterly at 1.5% for 30 years? Use the formula:

$$A = P\left(1 + \frac{APR}{n}\right)^{nY}$$

a) Identify the variables for the formula.
   - P (starting principal amount) =
   - APR (annual percentage rate of interest as a decimal) =
   - n (number of compounding periods per year) =
   - Y (number of years) =

b) Calculate the value of A.

   $A =$

(Do not round until the final answer. Then round to the nearest cent.)

2. How much would I need to deposit today into an account at 6% compounded monthly if I want to have $100,000 in 15 years? Use the formula:

$$P = \frac{A}{\left[\left(1 + \frac{APR}{n}\right)^{nY}\right]}$$

a) Identify the variables for the formula.
   - A (accumulated balance after Y years) =
   - APR (annual percentage rate of interest as a decimal) =
   - n (number of compounding periods per year) =
   - Y (number of years) =

b) Calculate the value of P.

   $P =$

(Do not round until the final answer. Then round to the nearest cent.)

3. Dave invests $5000 into an account at 3.5% compounded weekly for 4 years. Kim invests $5000 into a different account at 3% compounded continuously for 4 years. Who has the most money in their account after 4 years?

4. If Brian deposits $220 into an account each month with an *APR* of 4%. How much will he have accumulated in 10 years? Use the formula:

$$A = PMT \cdot \left[\frac{\left(1 + \dfrac{APR}{n}\right)^{nY} - 1}{\dfrac{APR}{n}} \right]$$

a) Identify the variables for the formula.
   - *PMT* (payment amount) =
   - *APR* (interest rate changed to a decimal) =
   - *n* (number of payments per year) =
   - *Y* (number of years) =

b) Calculate the value of *A*.

   $A =$

(Do not round until the final answer. Then round to the nearest cent.)

5. If Linda needs $8000 to buy a used car in 5 years, how much should she deposit weekly at an interest rate of 3.5%? Use the formula:

$$PMT = \frac{A \cdot \left(\dfrac{APR}{n} \right)}{\left[\left(1 + \dfrac{APR}{n}\right)^{(nY)} - 1 \right]}$$

a) Identify the variables for the formula.
   - *APR* (interest rate changed to a decimal) =
   - *n* (number of payments per year) =
   - *Y* (number of years) =
   - *A* (desired accumulated amount) =

b) Calculate the value of *PMT*.
   PMT (round to the nearest cent as needed) =

6. Patty has a mortgage of $349,500. What would her monthly payments be if her APR is 4.6% and her loan is for 15 years? Use the formula:

$$PMT = \frac{P \cdot \left(\frac{APR}{n}\right)}{\left[1 - \left(1 + \frac{APR}{n}\right)^{(-nY)}\right]}$$

a) Identify the variables for the formula.
   - APR (interest rate changed to a decimal) =
   - n (number of payments per year) =
   - Y (number of years) =
   - P (principle borrowed) =

b) Calculate the value of PMT.
   PMT (round to the nearest cent as needed) =

UNIT III
BASIC SKILLS FOR PROBABILITY AND STATISTICS

5.1 Calculating Relative Frequency as Fraction, Decimal, and Percentage

It is often necessary to examine a frequency table and be able to convert the frequency count in each class into a fraction of the total, a decimal part of the total or percentage.

Consider the frequency table given below:

Class	Frequency Count
0–9	42
10–19	58
20–29	50

Step 1: To find the fractional part represented by each class in the table, we need to find the total frequency count:

$$42 + 58 + 50 = 150$$

Step 2: The fraction represented by each class is the number in the class, divided by the total count.

Class	Frequency Count	Fraction
0–9	42	42/150
10–19	58	58/150

Class	Frequency Count	Fraction
20–29	50	50/150

Step 3: The decimal part is simply converting this fraction into a decimal.

Class	Frequency Count	Fraction	Decimal
0–9	42	42/150	0.2800
10–19	58	58/150	0.3867
20–29	50	50/150	0.3333

(Note that the decimal equivalent has been rounded to the fourth place.)

Step 4: Finally, the percent for each class is found by converting the decimal part into a percent.

Class	Frequency Count	Fraction	Decimal	Percentage
0–9	42	42/150	0.2800	28%
10–19	58	58/150	0.3867	38.67%
20–29	50	50/150	0.3333	33.33%

5.1 PRACTICE PROBLEMS

1. Use your knowledge of conversions among fractions, decimals, and percent to fill in the frequency tables in the practice problems.

Class	Frequency Count	Fraction	Decimal	Percentage
3–8	2			
9–14	12			
15–20	64			
21–26	18			
27–32	7			

2. The following table shows the frequency counts for 25 students on a statistics exam. Complete the conversions for the remainder of the table.

Grade Range	Count	Fraction	Decimal	Percentage
30–39	1			
40–49	1			
50–59	2			
60–69	7			
70–79	8			
80–89	4			
90–100	2			

3. Complete the table below.

Age Range	Count	Relative Frequency (as a fraction)	Relative Frequency (as a decimal)	Relative Frequency (as a percentage)
1–3	8			
4–6	6			
7–9	12			
10–12	11			
13–15	9			

4. Consider the frequency distribution for the percent of high school completion for the counties in Appalachia.

Class	Frequency
50.0–59.9	1
60.0–69.9	22
70.0–79.9	166
80.0–89.9	211
90.0–99.9	20

a) How many total counties are there are there in Appalachia?

b) Use the table above, convert it to a relative frequency table and record the results in the table below.

Class	Relative Frequency (decimal)	Relative Frequency (percentage)
50.0–59.9		
60.0–69.9		
70.0–79.9		
80.0–89.9		
90.0–99.9		

5.2 Interpreting Charts and Graphs

Being able to properly interpret graphical materials is critical to mathematical literacy and useful in everyday life. Charts and graphs are used frequently to convey information and persuade us to buy products, take specific positions, or vote for certain candidates, to give just a few examples. There are some general procedures you should apply whenever you examine a chart or graph.

1. Look at the graph carefully. Many deceptions are based on the fact that most people merely glance at graphics and use that first impression without really scrutinizing the evidence.
2. Be sure to analyze the horizontal and vertical axes when looking at a line graph or bar chart. Do the axes begin at zero? What are the units of measurement? How are the axes marked off?
3. Line graphs should be read from left to right just like the English language. They can be easily misunderstood if you don't examine them from left to right.
4. Are there additional artistic elements in the graph that are meant to draw the attention away from the content information?

Consider the following examples.

EXAMPLE 1

Histograms

A histogram, like the one shown below, is similar to a bar graph except that the bars have no separation between them. They are needed to begin analyzing quantitative data. A histogram can show you how the data is distributed and can be used to estimate the center of the data set. A word of caution: histograms can't be used to *precisely* determine the mean or other statistics.

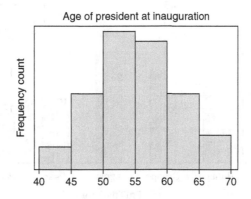

To get an estimate of the value of the mean, you can think about where a cardboard cut-out of the histogram would "balance" on a fingertip. It appears that the mean or average age of US Presidents at inauguration is about 55. We can't be much more accurate than this estimate. Further, it appears that the distribution of the ages is approximately normal (bell-shaped).

EXAMPLE 2

Line Graphs

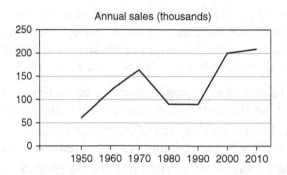

Looking at the graph it is clear that time is the horizontal axis which begins in 1950. The vertical axis appears to be annual sales in thousands of dollars and begins at zero. Reading from left to right, sales increased between 1950 and 1970 but then went down. Between 1980 and 1990 the sales were level (neither increasing nor decreasing). After 1990 the sales began to increase again, steeply increasing at first and then slowing down after 2000.

EXAMPLE 3

Bar Graphs

Bar graphs are useful when we want to compare the data characteristics of two or more categories at the same time. Consider the bar graph below:

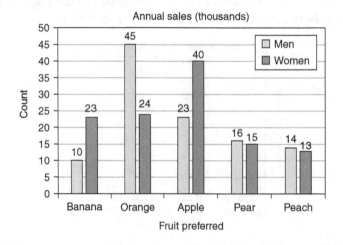

Here we are looking at the kind of fruit preferred by men and women. The bar graph allows us to compare the fruit preference of men and women on the same graph. Notice that this looks like a histogram, but the horizontal values are categorical (type of fruit) rather than the quantitative (numerical) data of a histogram.

EXAMPLE 4

Correlation comparisons: correlation is NOT cause and effect
We say that two variables are correlated to each other when they have a definable mathematical relationship. However, it is a very common error to think that a correlation between variables implies that one variable CAUSES a change in the other variable.

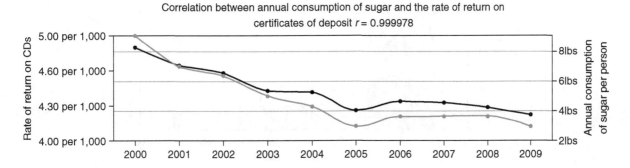

Correlation between annual consumption of sugar and the rate of return on certificates of deposit $r = 0.999978$

The rate of return on certificates of deposit does seem to correlate to the consumption of sugar. However, that doesn't mean that an increase in sugar consumption would cause a change in the rate of return on a bank CD. Of course, this sounds silly and in this case it is obvious that cause/effect doesn't exist. However, confusing correlation with causality is a common error that people looking at correlations make.

5.2 PRACTICE PROBLEMS

To answer Questions 1–3, use the histogram below:

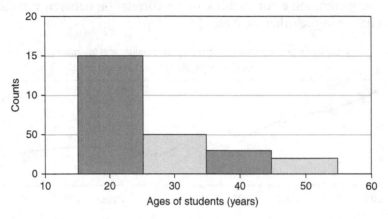

1. What appears to be the average age of the class?

2. How many students are above 35 years old?

3. How many students were in the class?

To answer Questions 4–6, use the bar chart below:

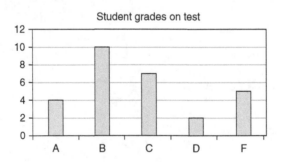

4. How many students passed the test with a C or above?

5. How many students failed the test?

6. Does this graph seem plausible?

To answer Questions 7–9 use the line graph below:

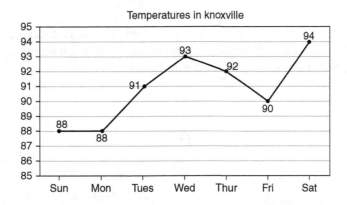

7. What is the mean temperature?

8. In what 24-hour period was there the greatest increase in temperature?

9. When did the temperature stay the same?

5.3 Plotting Points on a Cartesian Plane

To describe the location of points on a plane, we use a rectangular coordinate system which is called the Cartesian plane. This system is simply two number lines set perpendicular to each other with an intersection at the zero, called the *origin*, of each number line.

Cartesian Plane:

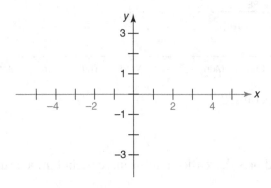

Plotting the location of a given ordered pair

Locations on the plane are represented by ordered pairs of numbers (x, y) where x indicates the position along the x-axis and y indicates the position along the y-axis. To plot points, begin at the center or origin for both x- and y-values.

EXAMPLE 1

Plot the location of the ordered pair $(3, -2)$:

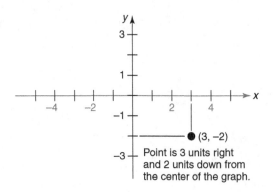

Point is 3 units right and 2 units down from the center of the graph.

Describing the location using an ordered pair

We can also examine graphs and describe the location using ordered pairs (x, y).

EXAMPLE 2

State the location of the points A, B, and C on the graph below.

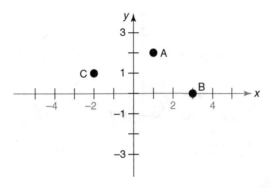

- Point A appears to be 1 unit right on the x-axis and 2 units up on the y-axis from the center of the graph $(0, 0)$. Therefore, the ordered pair is $(1, 2)$.
- Point B appears to be 3 units right and 0 units up from the center of the graph. Therefore, the ordered pair is $(3, 0)$.
- Point C appears to be 2 units left and 1 unit up from the center of the graph. Therefore, the ordered pair is $(-2, 1)$.

5.3 PRACTICE PROBLEMS

1. Identify the location of the labeled points:

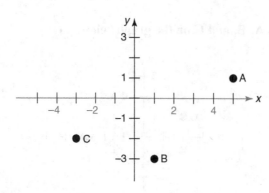

    a. Point A:
    b. Point B:
    c. Point C:

2. Plot and label the following ordered pairs on the Cartesian graph below:

Point A: $(0,-3)$ Point B: $(1,2)$ Point C: $(-1,-1)$ Point D: $(-4, 2)$

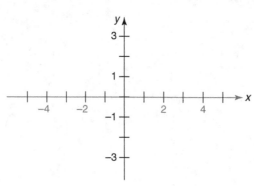

3. Plot and label the points from the following table.

x	y
−3	−4
−1	−2
0	−1
3	2

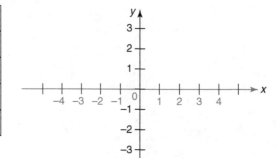

5.4 Order of Operations in Measures of Central Tendency

Here is a basic review of the order of operations.

Order of Operations Review

You will use the order of operations to perform algebraic computations. The order of operations are as follows:

Parentheses | Exponents | Multiplication or Division | Addition or Subtraction

1. Perform the operations inside parentheses and other grouping symbols such as brackets and division bars first.
2. Then exponents.
3. Then multiplication and division, from left to right.
4. Then addition and subtraction, from left to right.

The acronym PEMDAS is used as a mnemonic device to help remember the order of operations. Let's review the most basic rules of arithmetic operations.

Integer Addition

- Positive + Positive = Positive: $4 + 7 = 11$
- Negative + Negative = Negative: $(-5) + (-2) = -7$

When you have a problem with a positive value and a negative value, you need to subtract the absolute values of the two numbers and then the answer gets the sign of the number with the larger absolute value.

$$-11 + 2 = -9$$

Integer Subtraction

- Negative − Positive = Negative: $-7 - 6 = -7 + (-6) = -13$
- Positive − Negative = Positive + Positive = Positive: $7 - (-2) = 7 + 2 = 9$
- Negative − Negative = Negative + Positive.

Therefore, you need to consider the absolute values and proceed as for addition. $-2 - (-8) = -2 + 8 = 6$

Integer Multiplication

- Positive × Positive = Positive: $8 \cdot 2 = 16$
- Negative × Negative = Positive: $-4 \cdot (-3) = 12$
- Negative × Positive = Positive × Negative = Negative: $-8 \cdot 5 = 5 \cdot (-8) = -40$

Integer Division

- Positive ÷ Positive = Positive: $\frac{6}{2} = 3$
- Negative ÷ Negative = Positive: $\frac{-40}{-5} = 8$
- Negative ÷ Positive = Positive ÷ Negative = Negative: $\frac{15}{-3} = \frac{-15}{3} = -5$

Order of Operations and the Mean of a Sample

A common formula to consider is the formula for the mean of a sample.

$$\bar{x} = \frac{sum\ of\ all\ data\ values,\ x}{total\ number\ of\ values} = \frac{\Sigma x}{n}$$

Because of the division bar, the numerator operation should be done first. This is like putting the numerator in a parenthesis:

$$\bar{x} = \frac{(\Sigma x)}{n}$$

In this formula, you must also remember that the symbol Σ means that you must do a summation. Since the entire phrase is Σx, the directive is to sum all the values of x in the sample. Then you would finish by dividing by n, the total number in the data set.

EXAMPLE 1

Consider the calculation of the mean for the data set given below:

2, 6, 2, 4, 5

Step 1: Determine how many numbers are in the data set. Since there are five numbers in the data set, $n = 5$.

Step 2: Substituting into the formula for the mean:

$$\bar{x} = \frac{\Sigma x}{n} = \frac{(\Sigma x)}{n} = \frac{(2+6+2+4+5)}{5} = \frac{19}{5} = 3.8$$

- Note that the summation was done first as required by Step 1 of the order of operations. Step 2 was skipped because there are no exponents. Finally, the division of the numerator by the denominator satisfies Step 3 of the order of operations.

Order of Operations and the Midrange

Specifically, we need to use the order of operations correctly when applied to the midrange which is another, although less commonly used, measure of center. The formula for computing the midrange is as follows:

$$midrange = \frac{Maximum + Minimum}{2}$$

As noted earlier, the division bar means the numerator of the fraction is implied to be in a parenthesis. Thus:

$$midrange = \frac{(Maximum + Minimum)}{2}$$

EXAMPLE 2

Consider finding the midrange of the following data list:

$$5, 8, 4, 6, 14, 40, 6, 8, 10, 9$$

Step 1: We need to determine the maximum number and minimum number in the data set. We do this by sorting the data list from lowest to highest.

When sorted this data list is: 4, 5, 6, 6, 8, 8, 9, 10, 14, 40

So the maximum number is 40 and minimum is 4. Substituting into the midrange equation:

$$midrange = \frac{(Maximum + Minimum)}{2} = \frac{(40 + 4)}{2}$$

Step 2: Applying the order of operations, do the operation in the numerator.

$$midrange = \frac{(40 + 4)}{2} = \frac{44}{2}$$

Step 3: There are no exponents, so we next do the division.

$$midrange = \frac{(Maximum + Minimum)}{2} = \frac{(40 + 4)}{2} = \frac{44}{2} = 22$$

Order of Operations and the Median

The median is defined as the middlemost data value in an ordered data set. In cases where there is an odd number of data values, the median requires no calculation to determine. It is the middle number in the sorted data set.

In data sets with an even number of data values, there are two data values defining the middle of the sorted data. In this case, you must "average" these two numbers to determine the median. Consider the example below.

EXAMPLE 3

Find the median for the data set: $\{1, 3, 5, 8, 10, 14, 20, 22, 25, 32\}$

Step 1: There are 10 data values in this set. So, we need to find the two values in the center. These two numbers are 10 and 14.

Step 2: Average these two numbers by following the formula:

$$median = \frac{value\ 1 + value\ 2}{2}$$

Substituting:

$$median = \frac{value\ 1 + value\ 2}{2} = \frac{10 + 14}{2}$$

Step 3: The numerator is in an "implied" parenthesis. Therefore, do the numerator addition first.

$$median = \frac{10+14}{2} = \frac{24}{2}$$

Step 4: Complete the problem by dividing the numerator by the denominator.

$$median = \frac{24}{2} = 12$$

5.4 PRACTICE PROBLEMS

Perform the indicated operation for Problems 1–5:

1. $8 + 5 =$

2. $-3 + 2 =$

3. $-2 - 6 =$

4. $-5 \cdot 3 =$

5. $-7 \cdot (-4) =$

To answer Questions 6–8, use the following data set:

$$3, 8, 2, 5, 6, 5, 4, 8, 7, 3$$

6. Find the midrange value.

7. Find the mean.

8. Find the median.

5.5 Evaluating Rational Roots in Measures of Variation

Radicals

In evaluating the standard deviation by hand it is necessary to take a square root. The $\sqrt{}$ symbol is called the radical symbol. A square root is written as $\sqrt[2]{}$. In this case, the index number (which tells you which root you are seeking) is the number 2. Generally, however, we do not write the index number 2, instead assume that $\sqrt{} = \sqrt[2]{}$. A square root is also a fractional exponent with the index number being the denominator of the fractional exponent. Thus, $\sqrt{x} = x^{1/2}$.

However, if the index number is not a 2, then we must write it. For example, the cube (third) root is written as $\sqrt[3]{}$. The cube root is also a fractional exponent, since $\sqrt[3]{x} = x^{1/3}$. Notice that the index number is always the denominator of the fractional root.

When you use the radical symbol, you are asking yourself, "What number would I have to multiply by itself the index number of times to get the value under the radical symbol?"

EXAMPLE 1

Find $\sqrt[2]{36}$.

> To determine the answer, we need to ask what number could be multiplied by itself 2 times and get 36? The answer is 6, because $6 \cdot 6 = 36$.

EXAMPLE 2

Find $\sqrt[3]{125}$.

> To determine the answer, we need to ask what number could be multiplied by itself 3 times and get 125? The answer is 5, because $5 \cdot 5 \cdot 5 = 125$.

Of course, most roots do not turn out to be perfect whole numbers. This is why we use the calculator to do the evaluations for us and give us decimal approximations of the value of the root.

EXAMPLE 3

Use the calculator to find the square root of 18.

$$\sqrt{18} \approx 4.24.$$

> This means that $4.24 \cdot 4.24$ *is about* 18. Remember that we rounded the answer of 4.24, so it is not an exact square root.

EXAMPLE 4

Use the calculator to find $\sqrt[4]{200}$.

There is not a button on the calculator specifically for the fourth root. So, we use a more general procedure:

Step 1: Type in the index number of 4.

Step 2: Press the MATH button and go down to the function number 5: $\sqrt[x]{\ }$ and press enter.

Step 3: Now the window should show the calculation you want: $\sqrt[4]{\ }$ so all you need to do is type in the number 200 and press ENTER.

$$\sqrt[4]{200} = 3.76 \ rounded \ to \ the \ hundredth \ place$$

Standard Deviation

The formula for standard deviation is a more complex case of using the order of operations because it does contain an exponent (the square root). Let's look at the formula and break it down:

$$s = \sqrt{\frac{sum \ of \ (deviations \ from \ the \ mean)^2}{total \ number \ of \ data \ values - 1}} = \sqrt{\frac{\sum (x-\bar{x})^2}{n-1}} = \left(\frac{\sum (x-\bar{x})^2}{n-1}\right)^{1/2}$$

Remember that the square root is the same as raising the stuff under the square root symbol (the radicand) to the ½ power.

So, applying the order of operations, Step 1 means that we must evaluate the expression:

$$\frac{\sum (x-\bar{x})^2}{n-1}$$

before we can take the square root.

Considering that the expression $\frac{\sum (x-\bar{x})^2}{n-1}$ is a fraction, we know that we must do the operations in the numerator first. That is, we must evaluate $\sum (x-\bar{x})^2$ before we do anything else, but note that there is a parenthesis here also. So Step 1 of the order of operations forces us to find all the values of $(x-\bar{x})^2$ and then add them all together. Once we have done that summation, then we can divide by $n-1$. Finally, we will take the square root of the quotient, the answer obtained by the division of numerator by denominator.

Complicated, huh? An example using real numbers should help.

EXAMPLE 5

Consider the data set {2, 6, 2, 4, 5} and calculate the standard deviation.
Tabular form is being used because it makes the process easier.

Step 1: Determine the mean. Using the data set {2, 6, 2, 4, 5}, the mean $\bar{x}$ is 3.8.

Step 2: Subtract the mean from each data value then square the result.

x	$(x - \bar{x})$	$(x - \bar{x})^2$
2	$(2 - 3.8) = -1.8$	$(-1.8)^2 = 3.24$
6	$(6 - 3.8) = 2.2$	$(2.2)^2 = 4.84$
2	$(2 - 3.8) = -1.8$	$(-1.8)^2 = 3.24$
4	$(4 - 3.8) = 0.2$	$(0.2)^2 = 0.04$
5	$(5 - 3.8) = 1.2$	$(1.2)^2 = 1.44$

Notice that you need to use a parenthesis around the differences that are negative. Any number when squared is positive.

Step 3: Now we need to perform the operation in the numerator by finding the sum of those the values in the last column:

$\sum(x - \bar{x})^2$	$3.24 + 4.84 + 3.24 + 0.04 + 1.44 = 12.8$

Step 4: Next, perform the division under the radical sign or inside the parentheses.

$$s = \left(\frac{\sum(x - \bar{x})^2}{n-1}\right)^{1/2} = \left(\frac{12.8}{(n-1)}\right)^{1/2} = \left(\frac{12.8}{(5-1)}\right)^{1/2} = \left(\frac{12.8}{4}\right)^{1/2}$$

Then, because the square root is equivalent to raising the amount inside the square root sign to the one-half power, this is the same as saying:

$$s = \left(\frac{12.8}{4}\right)^{1/2} = (3.2)^{1/2}$$
$$= \sqrt{3.2} = 1.789$$

(rounded to three decimal places)

5.5 PRACTICE PROBLEMS

1. Find the roots indicated:
   a. $\sqrt{16}$

   b. $\sqrt{24}$

   c. $\sqrt[3]{125}$

   d. $\sqrt[4]{16}$

2. Express the root shown as a fractional exponent and solve.
   a. $\sqrt{5}$

   b. $\sqrt[5]{32}$

3. Find the value of the standard deviation using the order of operations. Show all intermediate steps.

$$s = \sqrt{\frac{6(26,870)-(324)^2}{6(6-1)}}$$

4. Find the value of the standard deviation. Show all intermediate steps.

$$s = \sqrt{\frac{6(105,965)-(795)^2}{6(6-1)}}$$

5. Find the standard deviation for the data set given. You need to calculate the mean before filling in the rest of table. Show all of your work.

$$\{1, 2, 3, 4, 5, 6\}$$

   a. Mean =

   b. Fill in the table below:

Data value	$(x - \bar{x})$	$(x - \bar{x})^2$
1		
2		
3		
4		
5		
6		

c. Sum of the (deviations from the mean)^2 =

d. $s = \sqrt{\dfrac{sum\ of\ (deviations\ from\ the\ mean)^2}{total\ number\ of\ data\ values - 1}} =$

5.6 Interpreting and Graphing Linear Inequalities

Inequality Symbols

An inequality is used when we don't know exactly what an expression is equal to. Instead of an equals sign, we use one of these symbols:

Symbol	Meaning	Common Meaning
>	Greater than	"more than," "over," "above," or "bigger"
<	Less than	"fewer than," "under," "below," or "smaller"
≥	Greater than or equal to	"at least" or "no fewer than"
≤	Less than or equal to	"at most" or "no more than"

Comparing Numbers Using Inequalities

Comparison of decimal numbers is also needed in probabilities. To decide which number is larger requires that you first make the decimals the same length. Then "read" the decimal as if it is a fraction. This will allow you to make the comparison more easily.

EXAMPLE 1

Which probability is the largest 0.043 or 0.0099?

Step 1: Make the decimals the same length by adding zeros to the shorter number.

The number 0.043 needs to be have four places to the right of the decimal point. Adding a zero on the far right gives 0.0430. Now both numbers are the same length.

Step 2: Convert them to fractions.

$0.0430 = \frac{430}{10,000}$ which can be read as "430 ten-thousandths."

$0.0099 = \frac{99}{10,000}$ which can be read as "99 ten-thousandths."

Therefore, the first number (0.043) is larger than the second number (0.0099) because there are more "ten-thousandths" in the first number.

$$0.043 > 0.0099$$

We could follow a different method for deciding which decimal number is larger. First, align the decimal places and then, working from left to right, compare the numbers in each place value position, until you reach a place where the numbers are different. The larger value in that place is the larger decimal number.

EXAMPLE 2

What number is larger: 0.26894 or 0.269?

$$0.269 > 0.26894$$

(This is true because there is a 9 in the thousandths place of 0.269 when 0.26894 has an 8 in this place.)

EXAMPLE 3

What is the probability that a woman's height is between 58 and 62 inches tall?

To express this probability as an inequality, we need to consider the meaning of the word "between." Specifically, does the phrase "between 58 and 62" include these numbers? Correctly interpreted, the phrase would not include 58 or 62, only the numbers that are literally between them. On a number line this phrase "between 58 and 62" looks like this:

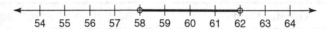

Written as an inequality this would be $P(58 < x < 62)$ where the value of x can only be numbers between, but not including, 58 and 62.

If you want to include the endpoints, 58 and 62, in the inequality, then it must be stated differently. If you ask "what is the probability that a woman has a height between 58 and 62 inches tall inclusive" then you do include the endpoints. The key word here is "inclusive" which means to include the endpoints in the interval. Now the number line would look like this:

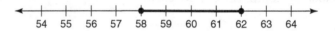

The inequality form would be $P(58 \leq x \leq 62)$. So x can take any value between 58 and 62, including the endpoints of 58 and 62.

EXAMPLE 4

What is the probability of getting a score below 70 on an exam?

This probability includes any test score below 70 but not 70 itself. Therefore, the number line representation would be:

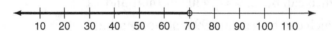

The inequality form would be $P(x < 70)$ which indicates that any score, x, below 70 is included.

EXAMPLE 5

What is the probability of getting at least a 70 on the exam?

This is different from the previous example because a score of 70 or anything higher would be included in this description. So the number line representation would be:

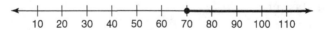

The inequality form would be $P(x \geq 70)$ because any score of 70 or over would fit this description.

5.6 PRACTICE PROBLEMS

For the following problems, interpret the probability statements as both number line graphs and as inequalities.

1. What is the probability that a student grade on a certain test lies between 80 and 92?

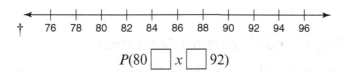

$$P(80 \boxed{} x \boxed{} 92)$$

2. What is the probability that an egg weighs 4 ounces or more?

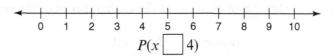

$$P(x \boxed{} 4)$$

3. What is the probability that an egg weighs more than 4 ounces?

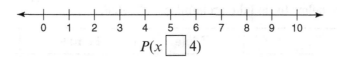

$$P(x \boxed{} 4)$$

4. What is the probability that a randomly selected person has an IQ below 130?

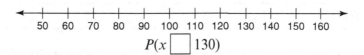

$$P(x \boxed{} 130)$$

5. What is the probability that a randomly selected person has an IQ between 100 and 130, inclusive?

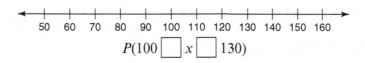

$$P(100 \boxed{} x \boxed{} 130)$$

5.7 Adding and Subtracting Fractions and the Addition Rule

When we perform operations using the addition rule for probability, we will need to both add and subtract fractions.

Procedure for Adding and Subtracting Fractions:

Step 1: Verify that all the denominators are the same. If they are not then we need to find a common denominator.

Step 2: Combine the arithmetic operations in the numerator over the common denominator.

Step 3: Solve the arithmetic operation in the numerator.

EXAMPLE 1

Use the contingency table below to find the probability indicated.

	Male	Female	TOTAL
Fears flying	7	5	12
Does not fear flying	3	20	23
TOTAL	10	25	35

If one person is selected at random from the table above, what is the probability of selecting a female or someone who does not fear flying?

The addition rule application would require us to solve this problem:

$$P(\text{female OR doesn't fear}) = P(\text{female}) + P(\text{doesn't fear}) - P(\text{female and doesn't fear})$$

We would get these individual fractional probabilities:

$$P(\text{female}) = \frac{25}{35}$$

$$P(\text{doesn't fear flying}) = \frac{23}{35}$$

$$P(\text{female and doesn't fear flying}) = \frac{20}{35}$$

Substituting into the addition rule formula:

$$P(\text{female OR doesn't fear flying}) = \frac{25}{35} + \frac{23}{35} - \frac{20}{35} =$$

All of the denominators are the same, so we can combine over the common denominator:

$$P(\text{female OR doesn't fear flying}) = \frac{25}{35} + \frac{23}{35} - \frac{20}{35} = \frac{25 + 23 - 20}{35} = \frac{28}{35}$$

EXAMPLE 2

A class has 23 people in it. Eleven are female and the rest are male. Fifteen people in the class are under 25 years of age and twelve of these are male. The rest of the class is 25 or older. If I select one person at random from the class list, what is the probability of selecting a male OR someone under 25?

The addition rule application would require us to solve this problem:

$$P(male\ or\ under\ 25) = P(male) + P(under\ 25) - P(male\ and\ under\ 25)$$

We would get these fractional probabilities:

$$P(male) = \tfrac{12}{23} \text{ , since 12 people in the class are men}$$

$$P(under\ 25) = \frac{15}{23}$$
$$P(male\ and\ under\ 25) = \frac{12}{23}$$

So, applying the addition rule:

$$P(male\ OR\ under\ 25) = \frac{12}{23} + \frac{15}{23} - \frac{12}{23}$$

All of the denominators are the same, so we can combine over the common denominator:

$$P(male\ OR\ under\ 25) = \frac{12 + 15 - 12}{23} = \frac{15}{23}$$

Complementary Events

To find the probability of the complement of any event, we need to subtract the fractional probability from 1. The mathematical logic for this is shown below:

$$P(event\ A) + P(not\ event\ A) = 1$$

Rearranging this equation using algebra:

$$P(not\ event\ A) = 1 - P(event\ A)$$

For some students this subtraction offers a challenge because they are subtracting a fraction from the whole number 1. When we subtract fractions we need to have a common denominator, which means we need to show the number 1 with the same denominator as the fractional probability. But this is easy if you remember that "any number (except 0) divided by itself equals one."

That is:

$$\frac{4}{4} = 1 = \frac{7}{7}$$

Procedure for Subtracting Fractions from 1:

Step 1: Change the 1 into a fraction with both the numerator and the denominator having the same number as the denominator of the fractional probability.

Step 2: Combine the arithmetic operation in the numerator over the common denominator.

Step 3: Solve the arithmetic operation in the numerator.

EXAMPLE 3

Let's say we know that the probability of winning some game of chance, P(win), is $\frac{3}{53}$. What is the probability of losing (not winning)?

Step 1: Recognize that losing is the complement of winning. So:

$$P(not\ win) = 1 - P(win) = 1 - \frac{3}{53}$$

Step 2: Convert the 1 into a fraction with the denominator of 53.

$$P(not\ win) = \frac{53}{53} - \frac{3}{53}$$

Step 3: Do the subtraction in the numerator.

$$P(not\ win) = \frac{53}{53} - \frac{3}{53} = \frac{53-3}{53} = \frac{50}{53}$$

5.7 PRACTICE PROBLEMS

Perform the indicated operation for Problems 1–4.

1. $\dfrac{3}{4} + \dfrac{7}{4} =$

2. $\dfrac{8}{9} - \dfrac{4}{9} =$

3. $\dfrac{1}{12} + \dfrac{2}{3} =$

4. $1 - \dfrac{15}{40} =$

5. In a box of 12 cupcakes, 4 are red velvet, 6 are salted caramel, and 2 are chocolate. If a cupcake is selected at random, the probability of choosing a red velvet or chocolate cupcake is

$$P(red\ velvet\ OR\ chocolate) = \frac{4}{12} + \frac{2}{12}$$

   Add the fractions to determine the total probability.

6. A coin and six-sided die are tossed at the same time. The probability of tossing either a 2 on the die or *heads* on the coin is given by the addition rule as:

$$P(2\ OR\ Heads) = \frac{1}{6} + \frac{1}{2}$$

   Find the overall probability by completing the addition.

7. Consider the contingency table below:

		Gender	
		Female	**Male**
Smoking Status	Smoker	39	26
	Nonsmoker	279	310

Find the probability of selecting a female or a nonsmoker if you select one name at random.

This probability is shown by the equation below:

$$P(female\ OR\ nonsmoker) = P(female) + P(nonsmoker) - P(female\ nonsmoker)$$
$$= \frac{318}{654} + \frac{589}{654} - \frac{279}{654}$$

Show your work to complete this problem.

8. Consider the contingency table below of counties in Appalachia by state and production of coal.

	AL	GA	KY	MD	SC	TN
Coal	10	0	19	2	0	3
No Coal	27	37	35	1	6	49
TOTAL	37	37	54	3	6	52

Find the probability of selecting a coal producing county or a county in Tennessee if you select one county at random.

This probability is shown by the equation below:

$$P(coal\ producing\ or\ TN\ county)$$
$$= P(coal\ producing) + P(TN\ county) - P(coal\ producing\ county\ in\ TN)$$
$$= \frac{34}{189} + \frac{52}{189} - \frac{3}{189}$$

Show your work to complete this problem.

9. Consider the contingency table below:

	Lieutenant Candidate	Captain Candidate
White	43	25
Black	19	8
Hispanic	15	8

Find the probability of selecting a Lieutenant candidate or a Hispanic if you select one person at random.

This probability is shown by the equation below:

$$P(Lieutenant\ OR\ Hispanic)$$
$$= P(Lieutenant) + P(Hispanic) - P(Lieutenant\ candidate\ who\ is\ Hispanic)$$
$$= \frac{77}{118} + \frac{23}{118} - \frac{15}{118}$$

Show your work to complete this problem.

10. If the probability of getting a defective engine part, $P(defective\ engine\ part) = \frac{2}{175}$, what is the probability of NOT getting a defective part?

11. If the probability of NOT having PB&J sandwich for lunch, $P(not\ PB\&J\ sandwich) = \frac{3}{5}$, what is the probability of having a PB&J sandwich?

UNIT IV

PRINCIPLES OF FUNCTIONS, LINEAR RELATIONSHIPS, AND GEOMETRY

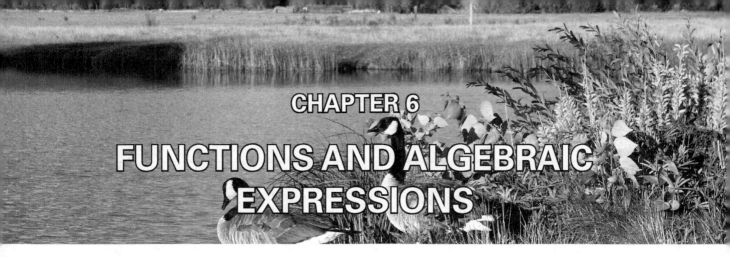

CHAPTER 6

FUNCTIONS AND ALGEBRAIC EXPRESSIONS

6.1 Determining Independent and Dependent Variables

In mathematics, we are often looking for a relationship between variables of interest. For example, we may theorize that grade level is related to reading comprehension. In that case, we may look for a linear relationship between these two variables.

Cartesian Plane:

It is customary in these analyses to consider one of the variables as the independent variable, plotted on the x-axis of the Cartesian plane, and define the other as the dependent or response variable, plotted on the y-axis.

Written description:

In a written description of a problem, it is customary to mention the independent variable first, followed by the dependent variable. So, in the example above, grade level would be considered the independent (x) variable, whereas reading comprehension would be the dependent (y) variable. In some cases, you need to consider the dependency relationship more carefully because this order is not ALWAYS used. (See Example 4 below.)

Tabular form:

When the data is presented in tabular form, the custom is for the x variable to be the first column in vertical presentations or first row in horizontal presentations.

EXAMPLE 1

Identify the independent and dependent variables in the problem below.

Data was collected to determine if there is a linear relationship between study time (hours) and tests scores.

Based upon the order of the variables in the description given:

Independent variable: study time

Dependent variable: test score

95

EXAMPLE 2

Identify the independent and dependent variables for the data set below:

Altitude (ft)	3000	8000	9000	15,000	20,000	25,000	33,000
Temperature(F)	62	55	22	0	−10	−20	−25

Based on the row order of the horizontal table:

Independent variable: altitude

Dependent variable: temperature

EXAMPLE 3

Identify the independent and dependent variables in the problem given.

Size of diamond (carats)	Price ($)
0.2	800
0.6	1100
1.6	5500
0.5	1200

Based on the column order of the vertical table:

Independent variable: size of diamond

Dependent variable: price

EXAMPLE 4

You and your friend go skydiving. How far have you descended since you jumped out of the plane?

In this case, the independent variable was not mentioned first in the statement. However, clearly the feet you have descended depends how long since you jumped.

Independent variable: time

Dependent variable: feet descended

6.1 PRACTICE PROBLEMS

1. At a routine visit to your child's pediatrician, she shows you a graph of age versus weight for children between 2 and 6 years old. Identify the independent and dependent variables in this problem.

   Independent variable:

   Dependent variable:

2. The rainfall for the month of June is between 0 and 3 in. The amount of water effects the quantity of blackberries.

   Independent variable:

   Dependent variable:

3. The time between eruptions of *Old Faithful* is related to the duration of the eruption.

   Independent variable:

   Dependent variable:

4. A pump-bottle of hand soap contains 12 ounces of fluid. Each pump extracts ¼ of an ounce of fluid. How much fluid remains in the bottle after a given number of pumps of the plunger?

   Independent variable:

   Dependent variable:

5. Examine the graph shown below and identify the independent and dependent variables.

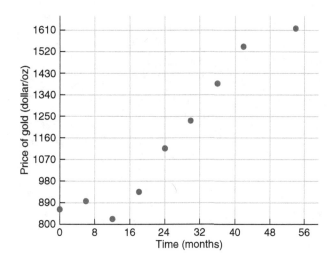

Independent variable:

Dependent variable:

6. Examine the data table given below and determine the independent and dependent variables.

Height (in.)	Arm Span (in.)
56	54
75	78
62	62
54	53
58	58

Independent variable:

Dependent variable:

7. Examine the data table given below and determine the independent and dependent variables.

Years of Education	8	12	20	14	16	18
Annual Salary (dollars)	24,520	25,200	52,026	34,525	48,000	48,200

Independent variable:

Dependent variable:

6.2 Interpreting Function Notation

Function notation:

We often develop short-hand ways of writing mathematical instructions. Function notation allows us to "name" a group of operations we want to perform on a specified independent variable. Consider the example below.

EXAMPLE 1

$$f(x) = 2x + 3$$

The function here is called f and the independent variable is x. You should not confuse this notation as indicating multiplication. In this short-hand notation, $f(x)$ is read as "f of x," which indicates that the function, f, has the independent variable x.

You could think of it like a mathematical factory: the independent variable is the input to the function (factory) and the output is called $f(x)$. Therefore, $f(x)$ is the dependent variable that we often call y.

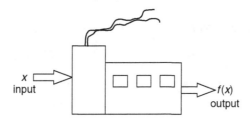

In the example above, the function process takes the value of x and multiplies it by 2 then adds 3 to get the output, $f(x)$.

Specifically, if $x = 2$ then $f(2) = 2 \cdot 2 + 3 = 7$.

Notice that $f(2)$ indicates that we wish to find the specific output value of the function when $x = 2$.

EXAMPLE 2

While it is common to name functions $f(x)$, you can use any other letter to name the function; and the independent variable can be other letters as well.

- $v(t)$ is often used for velocity functions that have the independent variable of time, t.
- $w(a)$ might indicate weight, w, as a function of age, a.

6.2 PRACTICE PROBLEMS

Answer the Questions 1–3 for the function below:

$$f(x) = 5x - 2$$

1. What is the independent variable?

2. What is the name of the function?

3. If $x = 1$, find $f(1)$. Show all your work.

Answer the Questions 4–6 for the function below:

$$s(t) = -16t + 200$$

4. What is the independent variable?

5. What is the name of the function?

6. If $t = 5$, find $s(5)$. Show all your work.

6.3 Evaluating Functions for Given Values

Evaluating for given values:

We are often asked to find the value of a function for a given value of the independent variable. The process is simply to substitute the value of the independent variable into the equation of the function and solve.

EXAMPLE 1

For the function, $f(x) = x^2 + 3x - 1$ find $f(2)$.

> **Step 1:** The instruction $f(2)$ directs us to use the value $x = 2$ in the function. Substituting this x value into the equation:
>
> $$f(2) = 2^2 + 3(2) - 1$$
>
> **Step 2:** Solve the equation using the order of operations.
>
> $$f(2) = 2^2 + 3(2) - 1 = 4 + 3(2) - 1 = 4 + 6 - 1 = 9$$

Table of Values:

To better understand how a function behaves over a set of values for the independent variable, we may be asked to create a table of values for the function. The values of the independent variable may be chosen by the students in some cases. However, the student often has to solve for values of the independent variable defined in a table.

Consider the example below.

EXAMPLE 2

The monthly cost of a certain long distance service is given by the linear function below:

$$C(t) = 0.07t + 3.95$$

where $C(t)$ is the time in minutes called in a month.

The independent variable in this equation is t, because the amount of the bill, $C(t)$, depends on how many minutes of long distance service is used. Find the cost of calling long distance for 80 minutes, 100 minutes, and 180 minutes, and complete the table.

Total Call Time, t (minutes)	Cost, $C(t)$
80	$C(t) = 0.07 \cdot 80 + 3.95 = 5.6 + 3.95 = 9.55$
100	$C(t) = 0.07 \cdot 100 + 3.95 = 7.0 + 3.95 = 10.95$
180	$C(t) = 0.07 \cdot 180 + 3.95 = 12.60 + 3.95 = 16.55$

So the completed table of values would look like:

t	C(t)
80	$9.55
100	$10.95
180	$16.55

6.3 PRACTICE PROBLEMS

For Problems 1–3, use the following function:

$$f(x) = x^2 - x + 3$$

1. Find $f(0)$.

2. Find $f(-1)$.

3. Find $f(2)$.

4. The price of gasoline is related to the cost of oil according to the equation $f(x) = 2.8x + 60.9$ where x is the cost of oil in dollars per barrel and $f(x)$ is the price of gasoline in cents per gallon. Evaluate the function for various values of x and fill in the table below:

Cost of Oil (dollars/barrel)	Price of Gasoline (cents/gallon)
20	
40	
60	
80	
100	
120	

5. The amount of ice cream sales in a local shop is related to the temperature outside by the equation $y = 20.5x - 956$ where y is the total sales (dollars) and x is the temperature in degrees Fahrenheit. Complete the table below:

Temperature (°F)	Total Ice Cream Sales ($)
58	
70	
54	
72	
60	

6. Forced expiratory volume, FEV, measures the amount of air (in liters) that a person can forcibly exhale in 1 second. There is a linear relationship between the height of a person and their FEV: $y = 0.13x - 5.43$ where x is height in inches and y is FEV in liters.

Height (in.)	FEV (liters)
52	
48	
62	
74	
46	

6.4 Solving Two-Step Equations

"Tips for solving"

Solving linear equations is the task of isolating the variable of interest, often x, and finding the value of this variable. To isolate the variable on one side of the equal sign, you must keep in mind the following:

- Start on the side with the variable.
- Look to do the "opposite" operation to remove any added or subtracted terms that are on this side.
- What you do to one side of the equal sign, you must do to the other side of the equal sign.
- Then look to do the "opposite" operation to undo any number that is multiplying or dividing the variable.

EXAMPLE 1

Solve for x in the equation: $5x - 1 = 19$

> **Step 1:** Addition and subtraction are operations that "undo" each other. To remove the subtracted 1 on the left side, we must add 1 to BOTH sides.

$$5x - 1 + 1 = 19 + 1$$
$$5x = 20$$

> **Step 2:** Multiplication and division are also operations that "undo" each other. To remove the 5 that is multiplying x, we must divide BOTH sides by 5.

$$\frac{5x}{5} = \frac{20}{5}$$
$$x = 4$$

EXAMPLE 2

Solve for x in the equation: $\frac{x}{2} + 5 = 35$

> **Step 1:** Eliminate the 5 by subtracting it from both sides.

$$\frac{x}{2} + 5 - 5 = 35 - 5$$
$$\frac{x}{2} = 30$$

> **Step 2:** Eliminate the 2 (which is currently dividing the x variable) by multiplying by it on both sides.

$$\frac{x}{2} \cdot 2 = 30 \cdot 2$$
$$x = 60$$

6.4 PRACTICE PROBLEMS

Solve problems 1–8 using the two-step process shown above. Show all of your work.

1. $3x + 7 = 19$

2. $\dfrac{x}{5} - 13 = 4$

3. $2x - 8 = 15$

4. $\dfrac{1}{2}x - 3 = 5$

5. $2z - \dfrac{2}{3} = \dfrac{7}{6}$

6. $-8 - 2x = -14$

7. $7x + 14 = 21$

8. $6 - z = 7$

6.5 Solving Multi-Step Equations

Only some linear equations in one variable can be solved with only two steps. Linear equations that have the variable on both sides of the equation, for example, require some additional steps. The additional steps often involve one or more of the following:

- Simplify both sides by combining like terms.
- Perform distribution when the variable is in parentheses.
- Use addition or subtraction to get the variable terms all on the same side of the equal sign and the constant terms on the other side.

Consider the following examples.

EXAMPLE 1

Solve for x in the equation: $4x - 3 = 22 - x$

Step 1: Since there is a variable term on both sides of the equation, we need to use addition (or subtraction) to move one of them to the opposite side. In this example, we will move the variable term from the right side to the left side by addition of an "x" to BOTH sides.

$$4x - 3 + x = 22 - x + x$$

Step 2: Combine the like terms on the left to further simplify.

$$5x - 3 = 22$$

Step 3: Remove the subtracted term on the left by addition.

$$5x - 3 + 3 = 22 + 3$$
$$5x = 25$$

Step 4: Remove the 5 that is multiplying the x value by division.

$$\frac{5x}{5} = \frac{25}{5}$$
$$x = 5$$

EXAMPLE 2

Solve for x in the equation: $6(x - 7) = 4$

Step 1: In this problem, we need to first distribute the 6 that is multiplying the parentheses.

$$6(x - 7) = 4$$
$$6x - 42 = 4$$

Step 2: Remove the subtracted term on the left by addition.

$$6x - 42 + 42 = 4 + 42$$
$$6x = 46$$

Step 3: Remove the 6 that is multiplying the x value by division.

$$\frac{6x}{6} = \frac{46}{6}$$

$$x = \frac{46}{6} = \frac{23}{3} = 7.67 \, (rounded)$$

EXAMPLE 3

Solve for x in the equation: $0.5(x + 4) = 2(3x - 1)$

Step 1: First distribute to simplify both sides.

$$0.5(x + 4) = 2(3x - 1)$$

$$0.5x + 2 = 6x - 2$$

Step 2: Move the variable term from the right side to the left side by subtraction of $6x$ on BOTH sides.

$$0.5x + 2 - 6x = 6x - 2 - 6x$$

Step 3: Simplify by adding like terms.

$$0.5x + 2 - 6x = 6x - 2 - 6x$$

$$-5.5x + 2 = -2$$

Step 4: Subtract 2 from both sides to eliminate it from the left side.

$$-5.5x + 2 - 2 = -2 - 2$$

$$-5.5x = -4$$

Step 5: Divide both sides by -5.5.

$$\frac{-5.5x}{-5.5} = \frac{-4}{-5.5}$$

$$x = 0.727 \, (rounded)$$

6.5 PRACTICE PROBLEMS

Solve for the variable. Show all of your work.

1. $-9r + 6 = 9r + 24$

2. $18 - 4x - 4 = 7x + 5x$

3. $-3(x + 2) = 12 + 3$

4. $\dfrac{3(r - 5)}{5} = 3r$

5. $3(x + 1) = -2(x - 3)$

6. $6 - 4a = a + 1$

7. $2x + 5 - x = 6x + 2$

8. $\dfrac{2(t - 3)}{5} = 3t$

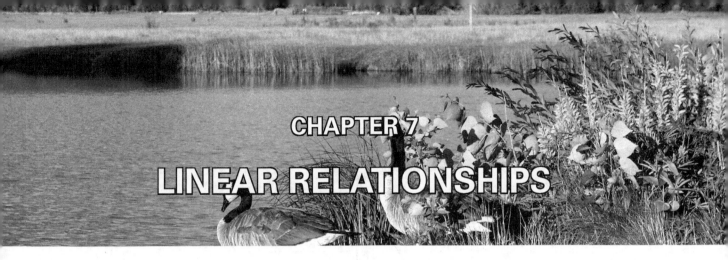

CHAPTER 7

LINEAR RELATIONSHIPS

7.1 Identifying Slope and Intercepts of a Linear Equation

A linear equation is used to describe the graph of a straight line in the Cartesian plane. There are three common forms for writing linear equations, depending on the information known about the graph.

Slope-intercept form:

When the slope and y-intercept are known, the form that we will find most useful is given by the literal equation below:

$$y = mx + b$$

In this form, x and y are the coordinates of any points that are on the line, m is the slope of the line, and b is the y-intercept.

By definition, slope describes the rate of change of the y values relative to the x values. When the line is going up (or increasing) from left to right, the line has a positive slope and m is a positive number. When the line is going down (or decreasing) from left to right, the line has a negative slope and m is a negative number.

The y-intercept represents the y-value of the point where the line crosses the vertical axis. This number can be either positive or negative. The x-coordinate of this point is always 0. Therefore, the ordered pair for the y-intercept can be written $(0,b)$.

In the slope-intercept equation below, the value of b is 3. This means that the line crosses the vertical axis at the coordinates $(0, 3)$. Examine the graph below as an illustration.

$$y = 2x + 3$$

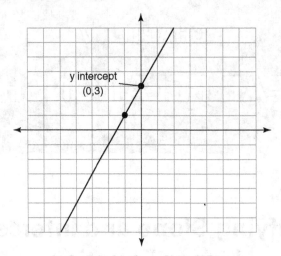

Let's look at some examples of how we would identify the slope and y-intercept for linear equations.

EXAMPLE 1

Identify the slope and y-intercept for the equation, $y = -2x + 5$

This equation is already in slope-intercept form so we simply inspect the equation. The slope is the coefficient of the x variable, so $m = -2$. The y-intercept is the constant term, $b = 5$. This can also be written as the ordered pair $(0,5)$.

If a linear equation is not written in slope-intercept form, you must first solve for y (by isolating the y variable) in order to identify the slope and y-intercept values. Once you have solved a linear equation for y, then the slope is the coefficient multiplying the x variable and the y-intercept is the constant term.

EXAMPLE 2

Identify the slope and y-intercept for the equation

$$3x - 3y = 6$$

Step 1: We need to put this equation into slope-intercept form by solving for y.

Subtract $3x$ from both sides.

$$3x - 3x - 3y = 6 - 3x$$
$$-3y = -3x + 6$$

Divide each term on both sides by -3.

$$\frac{-3y}{-3} = \frac{-3x}{-3} + \frac{6}{-3}$$
$$y = x - 2$$

Step 2: Identify the slope and the y-intercept. By inspection of the equation:

$$m = 1$$
$$b = -2 \text{ or } (0, -2)$$

EXAMPLE 3

Identify the slope and *y*-intercept for the equation

$$2x - y = 7$$

Step 1: Convert to slope-intercept form.

Subtract $2x$ from both sides.

$$2x - 2x - y = 7 - 2x$$
$$-y = 7 - 2x$$

Divide each term on both sides by -1.

$$\frac{-y}{-1} = \frac{7}{-1} - \frac{2x}{-1}$$
$$y = -7 + 2x$$

Step 2: Identify the slope, m, and the y-intercept. By inspection of the equation:

$$m = 2$$
$$b = -7 \text{ or } (0, -7)$$

7.1 PRACTICE PROBLEMS

Identify the slope and the intercept for each of the problems below. If necessary, solve the equation for y so that the equation is in slope-intercept form.

1. $y = -3x - 2$

2. $x = 2y - 3$

3. $y = x$

4. $y = 5x - 1$

5. $y = \dfrac{3}{2}x + 5$

6. $3x - 2 = y$

7. $2x + 4y = 13$

8. $14x - y + 2 = 0$

7.2 Interpreting Slope as Rate of Change

If you look at a graph of a line, the idea of slope is the amount of rise (vertical direction change) over the amount of run (horizontal direction change) between two points. The vertical direction change is the difference in the y-coordinates and the horizontal change is the difference in the x-coordinates.

We can express this relationship as an equation for the slope m between two points (x_1, y_1) and (x_2, y_2):

$$m = \frac{rise}{run} = \frac{y_2 - y_1}{x_2 - x_1}$$

Consider the examples below.

EXAMPLE 1

What is the slope of a line that contains the points (2, 14) and (5, 20)?

   Step 1: The slope formula is:

$$m = \frac{rise}{run} = \frac{y_2 - y_1}{x_2 - x_1}$$

   We need to identify the x- and y-values for each point.

   - Let the first point (2, 14) be (x_1, y_1) so that:

$$x_1 = 2 \text{ and } y_1 = 14.$$

   - Then the second point (5, 20) would be (x_2, y_2) so that:

$$x_2 = 5 \text{ and } y_2 = 20.$$

   Step 2: Substitute into the equation:

$$m = \frac{y_2 - y_1}{x_2 - x_1} = \frac{20 - 14}{5 - 2} = \frac{6}{3} = 2$$

In this particular case, the vertical direction change (or rise) is 2 and the horizontal direction change (or run) is an implied 1. This indicates that the line is "rising" twice as fast as it is "running."

A slope of 2 looks like this graphically:

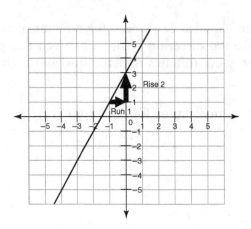

EXAMPLE 2

Calculate the slope of the line containing points (3, 8) and (5, 9).

   Step 1: The slope formula is:

$$m = \frac{rise}{run} = \frac{y_2 - y_1}{x_2 - x_1}$$

   • Let the first point (3, 8) be (x_1, y_1) so that:

$$x_1 = 3 \text{ and } y_1 = 8.$$

   • Then the second point (5, 9) would be (x_2, y_2) so that:

$$x_2 = 5 \text{ and } y_2 = 9.$$

   Step 2: Substitute into the equation:

$$m = \frac{y_2 - y_1}{x_2 - x_1} = \frac{9-8}{5-3} = \frac{1}{2}$$

Note that in this case, the line is changing vertically 1 unit for every horizontal change of 2 units. This slope would look like the following graph:

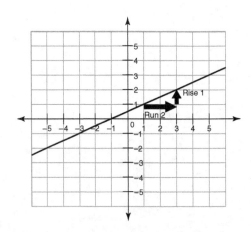

EXAMPLE 3

Calculate the slope of the line containing points (8, 14) and (11, 12).

Step 1: The slope formula is:

$$m = \frac{rise}{run} = \frac{y_2 - y_1}{x_2 - x_1}$$

- Let the first point (8, 14) be (x_1, y_1) so that:

$$x_1 = 8 \text{ and } y_1 = 14.$$

- Then the second point (11, 12) would be (x_2, y_2) so that:

$$x_2 = 11 \text{ and } y_2 = 12.$$

Step 2: Substitute into the equation:

$$m = \frac{y_2 - y_1}{x_2 - x_1} = \frac{12 - 14}{11 - 8} = \frac{-2}{3}$$

Note that in this case, the line has a vertical change of −2 (going down) for every horizontal change of 3. This slope would look like the following graph:

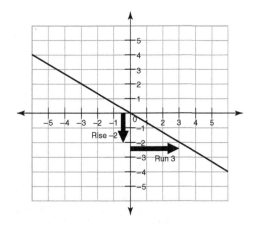

Note that as expected the line is falling from left to right, which is consistent with a negative slope value.

When a linear relationship is described in a word problem, the slope of the line can be interpreted as the rate of change, or the amount the dependent y-values change relative to the independent x-values.

Consider the example below.

EXAMPLE 4

I pay a flat fee of $4.99 per month and $0.25 per minute for international calls with my telephone service provider. This can be expressed as the linear equation:

$$y = 0.25x + 4.99$$

Describe what this means in terms of the rate of change of the dependent variable per unit change in the independent variable.

The dependent variable y is my total monthly cost and the independent variable x is the number of international minutes used that month.

The slope in this equation is 0.25, which tells us that my total monthly cost increases by \$0.25 for every international minute used.

7.2 PRACTICE PROBLEMS

For Problem 1–4, determine the slope as defined by the equation:

$$m = \frac{y_2 - y_1}{x_2 - x_1}$$

1. Point 1: (0,3)

    Point 2: (5,2)

    $$y_2 - y_1 =$$
    $$x_2 - x_1 =$$
    $$m =$$

2. Point 1: (−1,2)

    Point 2: (−6,4)

3. Point 1: (1,1)

    Point 2: (4,4)

4. Point 1: (5,0)

    Point 2: (5,1)

5. The cost, C, of a medium cup of frozen yogurt is \$3.25 plus \$0.30 for each additional topping, x, added. This can be represented by the linear equation:

    $$C = 3.25 + 0.30x$$

    a. What are the independent and dependent variables?

    b. What is the minimum cost (y-intercept) for a medium cup of frozen yogurt?

    c. Describe what the slope means in terms of the rate of change of the dependent variable per unit change in the independent variable.

6. The number of students enrolled at a community college each year after 2009 can be modeled by the linear equation:

$$E = 12,500 - 100t$$

a. What are the independent and dependent variables?

b. What is the maximum enrollment (y-intercept) of students at the college?

c. Describe what the slope means in terms of the rate of change of the dependent variable per unit change in the independent variable.

7.3 Creating a Table of Values from a Linear Equation

To create a table of values from a known linear relationship, we need to solve the linear equation that describes the relationship for several values of the independent or dependent variable. These values of these variables may be given or chosen on your own in some cases.

Consider the example below.

EXAMPLE 1

The cost of flawless diamonds less than 2 carats is related to their carat weight by the linear equation below:

$$C = -840 + 4950w$$

where C is the cost in dollars and w is the diamond weight in carats.

The independent variable in this equation is w, because the cost, C, depends on what size diamond you wish to purchase. Find the cost of buying flawless diamonds of carat weights 0.5, 1, and 1.5 carats, and complete the table.

Weight, w (Carats)	Cost, C (dollars)
0.5	$C = -840 + 4950(0.5) = \$1635$
1	$C = -840 + 4950(1) = \$4110$
1.5	$C = -840 + 4950(1.5) = \$6585$

So the completed table of values would look like:

Weight, w	Cost, C
0.5	$1635
1	$4110
1.5	$6585

EXAMPLE 2

Create a table of values and indicate the independent and dependent variables for the following:

A car's value decreases every year that it is driven. The Powell Motors Company briefly manufactured The Homer. The value of The Homer is given by the linear function.

$$V = -10,250t + 82,000$$

where t is the time in years and V is the value of the car.

Find the value of The Homer after 2 years, 5 years, and 8 years.

Based upon the description of the relationship:

Independent variable: time

Dependent variable: value of car

Time, t	2	5	8
Value, V	$V = -10,250 \cdot 2 + 82,000$ $= \$61,500$	$V = -10,250 \cdot 5 + 82,000$ $= \$30,750$	$V = -10,250 \cdot 8 + 82,000$ $= \$0$

EXAMPLE 3

Suppose that I have \$7000 to spend on a flawless diamond like those in example #1. What carat weight of diamond could I expect to buy?

Step 1: The equation relating the carat weight to the cost is given by:

$$C = -840 + 4950w$$

Substitute \$7000 into the equation:

$$7000 = -840 + 4950w$$

Step 2: Solve the equation for the weight w using the methods learned in Chapter 6.

$$7000 + 840 = -840 + 840 + 4950w$$
$$7840 = 4950w$$
$$\frac{7840}{4950} = \frac{4950}{4950}w$$

$$w = 1.6 \; carats \; (rounded \; to \; the \; nearest \; tenth \; of \; a \; carat)$$

7.3 PRACTICE PROBLEMS

Create a table of values for the following linear relationships:

1. $y = 3x - 2$

x	y
0	
−1	
2	

2. $2x + 3y = 6$

x	y
−3	
	−2
1	

3. $P = -150c + 10,500$

c	P
	9000
70	
	7500

4. Your car gets 30 mpg (that is 1/30 of a gallon per mile) and holds 16 gallons. How many gallons of gas are left in your tank if you drive 60, 90, or 180 miles?

The variable G represents the number of gallons of gas and represents the miles driven.

$$G = 16 - \frac{1}{30}m$$

m	G
60	
90	
180	

5. At one garment factory, it costs $20 to make each jacket and $350 to set up the production. What would be the cost of making 50, 60, or 100 jackets? The linear relationship between total cost, C, and number of jackets produced, j, is given by:

$$C = 20j + 350$$

j	C
50	
60	
100	

7.4 Graphing a Linear Equation Using the Slope and *y*-intercept

While we sometimes graph straight lines by creating a table of values and plotting individual points, we often find it easier to simply work from the linear equation in slope-intercept form, $y = mx + b$.

Consider the following examples.

EXAMPLE 1

Graph the equation $y = 2x + 3$.

> **Step 1:** We begin the graph at the *y*-intercept, the place where the line crosses the *y*-axis. The *x*-value here would have to be 0. The *y*-value of this point comes from the value of *b* in the equation, which is 3. In this case, the *y*-intercept written as a coordinate point is (0, 3).

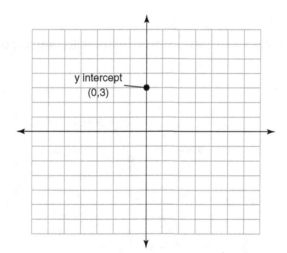

Notice that the *y*-intercept is a point on the graph of the line. The slope, *m*, is not a point, but a rate of change that shows how the *y* values change relative to the *x* values.

Step 2: From this point, we need to use the slope value to move to another point on the line.

$$m = \frac{rise}{run} = \frac{2}{1}$$

Therefore, we need to rise positive 2 units from our starting point while running positive 1 unit over.

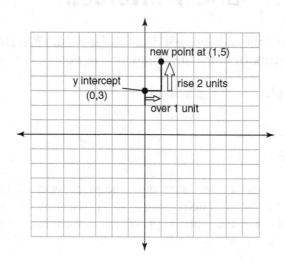

Step 3: Connecting these points with a straight line will complete our graph of the line $y = 2x + 3$.

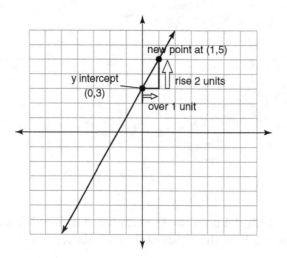

EXAMPLE 2

Graph the equation $2x + y = 4$.

Step 1: First, convert to slope-intercept form by solving for y.

$$2x + y - 2x = 4 - 2x$$
$$y = -2x + 4$$

Step 2: Identify the slope and the y-intercept. By inspection of the equation:

$$m = -2 = \frac{-2}{1} = \frac{rise}{run}$$
$$b = 4 \text{ or } (0, 4)$$

Step 3: Plot the *y*-intercept.

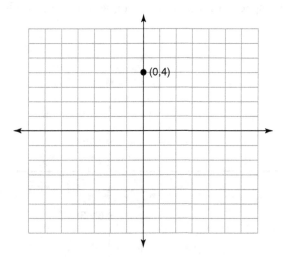

Step 4: Use the slope to move to another point on this line. Since the slope is −2, we will go down (the negative direction) by two and then over positive one.

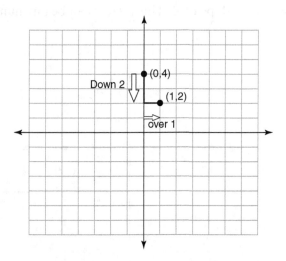

Step 5: Connect these points with a line and extend in both directions.

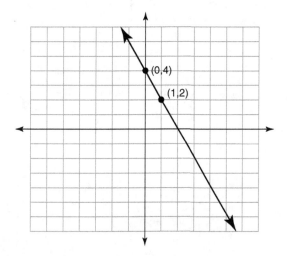

EXAMPLE 3

Vertical and horizontal lines are special cases. A vertical line like $x = 2$ can have any y value. It looks like this:

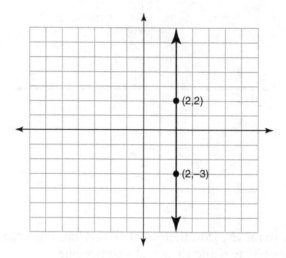

A horizontal line like $y = 3$ has a slope of 0. The x value can be any number.

It looks like this:

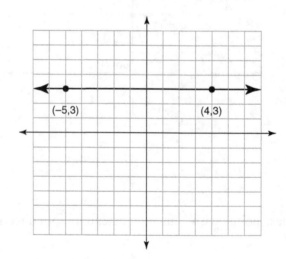

7.4 PRACTICE PROBLEMS

For Problems 1–4, use the graph below to draw each linear equation. Label each line and use a different color pencil for each one.

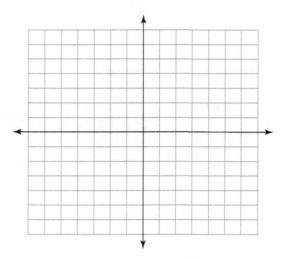

1. $y = -3x + 1$

2. $y = -1 + \dfrac{1}{2}x$

3. $y = \dfrac{1}{3}x + 2$

4. $y = 4$

For Problems 5–9, use the graph below to draw each linear equation. Label each line and use a different color pencil for each one.

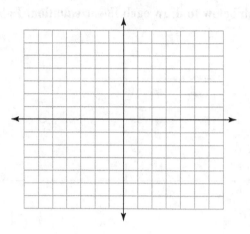

5. $x + 2y = 6$

6. $4x - 2y = 10$

7. $2x = 3 + 6y$

8. $x = -5$

9. $y = -\dfrac{2}{3}x + 5$

CHAPTER 8

GEOMETRY

8.1 Solving Problems Involving Perimeter and Circumference

Perimeter:

Perimeter is simply the length around the outside of any figure. In geometry, we usually deal with polygons, closed figures with straight sides such as rectangles and triangles. The drawing below shows the relationship between the lengths of the sides and the perimeter for rectangles and triangles.

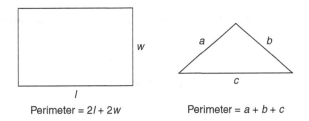

Perimeter = $2l + 2w$ Perimeter = $a + b + c$

The perimeter of a circle has a special name: circumference. The circumference of the circle depends on the length of its radius, which is defined as the distance between the center of the circle and any point on the circle. The circumference can also be defined using the diameter of the circle. Diameter is defined as the distance from a point on the circle to the point directly opposite it on the circle. The drawing below shows the radius and the diameter of the circle. Note that the diameter is twice the length of the radius. The number *pi* (π) is the ratio between the circumference of a circle and its diameter. It is a never-ending, never-repeating (irrational) number. We often estimate the value of *pi* by using 3.14, but this is not the exact value of *pi*. If we use this value as an estimate of *pi* to determine our answer, then the answer is also an estimate.

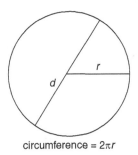

circumference = $2\pi r$

EXAMPLE 1

Determine the perimeter of a rectangle that has a length of 3 meters and a width of 2 meters.

Step 1: Identify the variables.

$$l = 3 \text{ meters}$$

$$w = 2 \text{ meters}$$

Step 2: Substitute these values into the formula for the perimeter of a rectangle.

$$P = 2l + 2w = 2 \cdot 3 + 2 \cdot 2$$

Step 3: Solve using the order of operations.

$$P = 2 \cdot 3 + 2 \cdot 2 = 6 + 4 = 10 \text{ meters}$$

Notice that the final units are meters because both of the dimensions were measured in meters. The dimension measurements should always be in the same units of measurement.

EXAMPLE 2

Determine the circumference of a circle whose diameter is 10 in. Leave the answer in terms of *pi*.

Step 1: Identify the variables.

$$d = 10 \text{ in.}$$

$$r = \frac{d}{2} = \frac{10 \text{ in.}}{2} = 5 \text{ in.}$$

Step 2: Substitute these values into the formula for the circumference of a circle.

$$C = 2\pi r = 2\pi (5 \text{ in.}) = 10\pi \text{ in.}$$

If required, we can estimate the circumference by using 3.14 as an estimate of the value of π.

$$C = 10\pi = 10(3.14) = 31.4 \text{ in.}$$

8.1 PRACTICE PROBLEMS

For the figures below, determine the perimeter or circumference:

1.

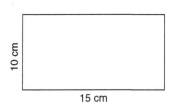

Perimeter =

2.

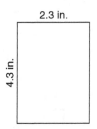

Perimeter =

3.

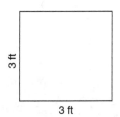

Perimeter =

4.

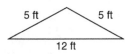

Perimeter =

5.

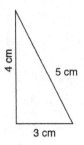

Perimeter =

6.

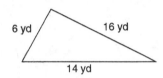

Perimeter =

7.

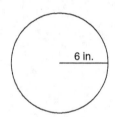

Circumference =

8.

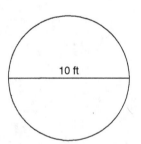

Circumference =

9. How much fencing will Tom need to surround his rectangular yard if the yard measures 100 ft long and 50 ft wide?

8.2 Determining Areas

Area:

As a concept, area describes the size of a two-dimensional surface. Calculating the area of a room, for example, could help us determine how many square yards of carpeting we need to purchase to cover the floor. There are several geometrical figures for which we need to be able to find the area.

Rectangles:

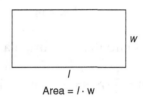

Area = $l \cdot$ w

EXAMPLE 1

Determine the area needed to fully carpet a room that is 10 ft by 12 ft

The rectangular area formula is needed. Substitute the known dimensions into the formula below. It doesn't matter which dimension is considered length and which is considered width. They can be interchanged.

> **Step 1:** The formula needed is $A = l \cdot w$

> **Step 2:** The problem description tells us the following:

$$l = 10 \text{ ft}$$

$$w = 12 \text{ ft}$$

> **Step 3:** Substitute these values into the area formula and solve using the order of operations.

$$A = l \cdot w = 10 \text{ ft.} \cdot 12 \text{ ft.} = 120 \text{ ft}^2$$

Notice that the final units are "square feet" because we multiplied feet times feet.

Triangles:

The formula for the area of a triangle is ½ times the base of the triangle times its height where the height must be perpendicular to the base. We often draw the picture:

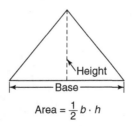

Area = $\frac{1}{2} b \cdot h$

EXAMPLE 2

Consider a triangle whose base is 7 in. and whose height is 2 in.

> **Step 1:** The formula needed is $A = \frac{1}{2} \cdot b \cdot h.$

> **Step 2:** The problem description tells us the following:

$$b = 7 \text{ in.}$$

$$h = 2 \text{ in.}$$

> **Step 3:** Substitute these values into the area formula and solve using the order of operations.

$$\text{Area} = \frac{1}{2} \cdot b \cdot h = \frac{1}{2} \cdot 7 \text{ in.} \cdot 2 \text{ in.} = 7 \text{ in.}^2$$

Notice that again the final units are squared.

EXAMPLE 3

Consider the large triangle shown below.

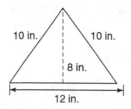

By examination, the 12 in. side is perpendicular (at right angles) to the 8 in. line. The 8 in. line is the height of the triangle even though it is not an actual side of the triangle.

> **Step 1:** The formula needed is $A = \frac{1}{2} \cdot b \cdot h.$

> **Step 2:** The problem description tells us the following:

$$b = 12 \text{ in.}$$

$$h = 8 \text{ in.}$$

> **Step 3:** Substitute these values into the area formula and solve using the order of operations.

$$\text{Area} = \frac{1}{2} \cdot b \cdot h = \frac{1}{2} \cdot 12 \text{ in.} \cdot 8 \text{ in.} = 48 \text{ in.}^2$$

Circles:

For circles, the area is defined by this formula: Area $= \pi r^2$ where the transcendental number, π, is often estimated by the number 3.14 and r is the radius of the circle. Remember that the radius of a circle is one-half of its diameter.

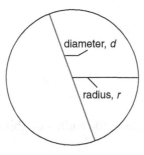

EXAMPLE 4

Consider the circle below:

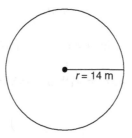

Step 1: The formula needed is $A = \pi r^2$.

Step 2: The problem description tells us the following:

$$r = 14 \text{ m}$$

Step 3: Substitute these values into the area formula and solve using the order of operations.

$$\text{Area} = \pi r^2 = \pi (14 \text{ m})^2 = 196 \ \pi \approx 615.44 \text{ m}^2$$

Again, notice that the units are square meters here. Also note that 615.44 m^2 is an approximation because we have used an estimate for the value of π.

EXAMPLE 5

Find the area of the circle shown below. Do not approximate π in your answer.

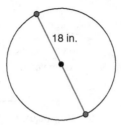

The radius is half of the diameter. Therefore, the radius of this circle is 9 in.

Step 1: The formula needed is $A = \pi r^2$.

Step 2: The problem description tells us the following:

$$r = 9 \text{ in.}$$

Step 3: Substitute these values into the area formula and solve using the order of operations.

$$\textbf{Area} = \pi r^2 = \pi (9 \text{ in.})^2 = 81\pi \text{ in.}^2$$

8.2 PRACTICE PROBLEMS

Find the area for the following figures:

1.

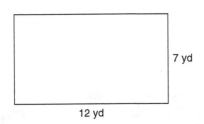

7 yd

12 yd

Area =

2.

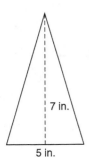

7 in.

5 in.

Area =

3.

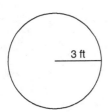

3 ft

Area =

4.

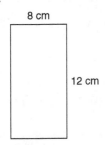

8 cm

12 cm

Area =

5.

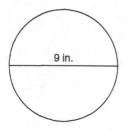

9 in.

Area =

6.

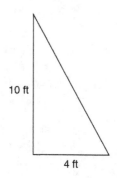

10 ft

4 ft

Area =

7.

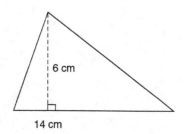

6 cm

14 cm

Area =

8. How much mulch would Nancy need to put in her rectangular garden that is 10 ft long and 3 ft wide?

8.3 Determining Volumes

Volume:

Volume is the amount of space within a three-dimensional object. We are able to determine the volumes of many common objects such as the rectangular solid, cylinder, and sphere. The volume formulas for these objects are given below:

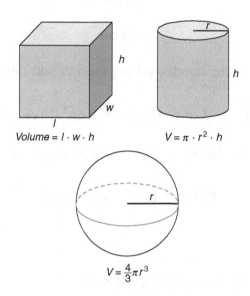

Volume = $l \cdot w \cdot h$ $V = \pi \cdot r^2 \cdot h$

$V = \frac{4}{3}\pi r^3$

Let's consider a few examples using these formulas.

Determine the volume of a child's building block that is a rectangular solid whose length is 6 in., width is 4 in., and height is 4 in.

    Step 1: The formula needed is $V = l \cdot w \cdot h$.

    Step 2: The problem description tells us the following:

$$l = 6 \text{ in.}$$

$$w = 4 \text{ in.}$$

$$h = 4 \text{ in.}$$

    Step 3: Substitute these values into the volume formula and solve using the order of operations.

$$V = l \cdot w \cdot h = (6 \text{ in.})(4 \text{ in.})(4 \text{ in.}) = 96 \text{ in.}^3$$

Notice that the final units are cubic inches because in. $\cdot$ in. $\cdot$ in. = in.^{3}.

EXAMPLE 2

What is the volume of a can of soup that is cylindrical with a radius of 3 cm and a height of 10 cm? Do not estimate π in the answer.

Step 1: The formula needed is $V = \pi r^2 h$.

Step 2: From the problem description, we know that:

$$r = 3 \text{ cm}$$

$$h = 10 \text{ cm}$$

Step 3: Substitute these values into the volume formula and solve.

$$V = \pi r^2 h = \pi \cdot (3 \text{ cm})^2 \cdot 10 \text{ cm} = \pi \cdot 9 \text{ cm}^2 \cdot 10 \text{ cm} = 90\pi \text{cm}^3$$

In some cases, we are asked to use the value 3.14 as an estimate for π when solving. In that case, the answer would be:

$$V \approx 90(3.14) \text{cm}^3 = 282.6 \text{ cm}^3$$

EXAMPLE 3

Find the volume of a soap bubble whose radius is 1.5 in.

Step 1: The formula needed is $V = \frac{4}{3}\pi r^3$.

Step 2: From the problem description, we know that:

$$r = 1.5 \text{ in.}$$

Step 3: Substitute these values into the volume formula and solve.

$$V = \frac{4}{3}\pi r^3 = \frac{4}{3}\pi (1.5 \text{ in.})^3 = \frac{4}{3}\pi \cdot 3.375 \text{ in.}^3 = 4.5\pi \text{ in.}^3$$

Using 3.14 to estimate π, the answer is:

$$V \approx 14.13 \text{ in.}^3$$

8.3 PRACTICE PROBLEMS

Find the volumes for the following problems:

1.

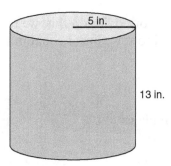

5 in.

13 in.

Volume =

2.

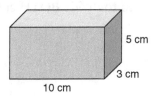

5 cm

3 cm

10 cm

Volume =

3. What is the volume of a sphere that has a radius of 7 ft?

4. How much does Emily's refrigerator hold if it has a height of 5 ft, width of 3 ft, and length of 4 ft?

5. What is the volume of a sphere that has a diameter of 10 in.?

6. How much would a cylindrical can hold that has a height of 18 cm and a radius of 7 cm?

7. How many cubic feet of water would Tim need to fill an aboveground circular swimming pool that is 16 ft across and 4.5 ft deep?

8. Billy wants to fill up his tub with Jello®. If the tub is 4 ft long, 1.5 ft wide, and 1.25 ft deep, how many cubic feet of Jello® will he need?

8.4 Solving for Given Variables in Literal Equations

Literal Equations:

Formulas and literal equations can be solved for any of the variables involved. To do this, we treat the variables as if they are "just numbers" and use the same steps as we used to solve linear equations. Consider the following examples.

EXAMPLE 1

In the formula for the area of a triangle, solve for the base, b. The formula is:

$$A = \frac{1}{2}bh$$

where A is the area, b is the base, and h is the height.

> **Step 1:** We wish to solve for the base, b, so we need to isolate it to one side. Right now, the base is being multiplied by ½ and by h. To undo these operations, we need to divide BOTH sides by them.

$$\frac{A}{\frac{1}{2}h} = \frac{\frac{1}{2}bh}{\frac{1}{2}h}$$

> **Step 2:** Simplifying by cancellation:

$$\frac{A}{\frac{1}{2}h} = \frac{\frac{1}{2}b\cancel{h}}{\frac{1}{2}\cancel{h}}$$

$$\frac{2A}{h} = b$$

So, we can find the base by taking the area, multiplying by 2 and dividing by the height.

EXAMPLE 2

Solve the cylindrical volume formula for the height of the cylinder. The formula for cylindrical volume is:

$$V = \pi r^2 h$$

where r is the radius, h is the height, and V is the volume.

Step 1: To isolate the requested variable, h, we note that it is being multiplied by πr^2. Therefore, we need to divide both sides of the formula by πr^2.

$$\frac{V}{\pi r^2} = \frac{\pi r^2 h}{\pi r^2}$$

Step 2: Simplifying by cancellation:

$$h = \frac{V}{\pi r^2}$$

EXAMPLE 3

Solve the formula for the perimeter of a rectangle for the length, l.

The formula for the perimeter of a rectangle is $\boldsymbol{P = 2l + 2w}$ where P is perimeter, l is length, and w is width.

Step 1: To isolate the requested variable, $\boldsymbol{l}$, we note that $2w$ is being added to the same side. Therefore, we need to begin by subtracting this term from both sides.

$$P - 2w = 2l + 2w - 2w$$

Step 2: Simplifying:

$$P - 2w = 2l$$

Step 3: Since the length, $\boldsymbol{l}$, is being multiplied by 2, we need to divide both sides by 2.

$$\frac{P - 2w}{2} = \frac{2l}{2}$$

Step 4: Simplifying by cancellation:

$$l = \frac{P - 2w}{2} = \frac{P}{2} - w$$

8.4 PRACTICE PROBLEMS

Solve for the indicated variable:

1. In the literal equation $V = l \cdot w \cdot h$, solve for w.

2. In the literal equation $P = 2l + 2w$, solve for w.

3. In the literal equation $y = mx + b$, solve for x.

4. In the literal equation $A = \pi r^2$ solve for π.

5. In the literal equation $E = mc^2$, solve for c.

6. In the literal equation $A = \frac{1}{2}(a + b)h$, solve for h.

7. In the literal equation $S = 2\pi rh + 2\pi r^2$, solve for h.

8.5 Solving Proportions

Proportions:

A proportion is an equation that states that two ratios (fractions) are equal to each other. If one term of a proportion is not known (usually designated by the variable x), the concept of cross multiplication can be used to find the value of the unknown term.

Here is an example of a proportion:

$$\frac{x}{3} = \frac{2}{9}$$

Remember that the numerator is the top number in a fraction and the denominator is the bottom number of the fraction. For example, in the fraction ½, 1 is the numerator and 2 is the denominator.

Cross Multiplication:

To cross multiply, you take the left side fraction's numerator and multiply it by the right side fraction's denominator. Then take the left side fraction's denominator and multiply it by the right side fraction's numerator. These two products are set equal to one another. Now you have eliminated the fractions and can solve as you would for other equations.

Consider the examples below.

EXAMPLE 1

Solve the proportion below for x.

$$\frac{x}{3} = \frac{2}{9}$$

Step 1: Cross multiply.

$$\frac{x}{3} \diagdown\!\!\!\!\diagup \frac{2}{9}$$

$$9x = 3 \cdot 2 = 6$$

Step 2: Divide both sides by 9.

$$\frac{9x}{9} = \frac{6}{9}$$

$$x = \frac{6}{9} = \frac{2}{3}$$

EXAMPLE 2
Solve the proportion for x:

$$\frac{4}{12} = \frac{9}{x}$$

Step 1: Cross multiply:

$$4x = 12 \cdot 9 = 108$$

Step 2: Divide both sides by 4.

$$\frac{4x}{4} = \frac{108}{4}$$
$$x = 27$$

EXAMPLE 3
Solve the proportion for x:

$$\frac{3}{8} = \frac{6}{x+4}$$

Step 1: Cross multiply:

$$3(x+4) = 8 \cdot 6 = 48$$

Note that the 3 is multiplying the entire quantity $(x+4)$.

Step 2: Distribute:

$$3x + 12 = 48$$

Step 3: Subtract 12 from both sides.

$$3x + 12 - 12 = 48 - 12$$
$$3x = 36$$

Step 4: Divide both sides by 3.

$$\frac{3x}{3} = \frac{36}{3}$$
$$x = 12$$

EXAMPLE 4

Proportions are used to solve for missing sides in similar triangles. Similar triangles have equal corresponding angles but unequal sides. However, the corresponding sides remain in the same ratio: they are proportional.

Consider the similar triangles below:

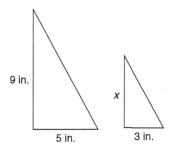

Step 1: Identify the corresponding sides.

The 9-in. side corresponds to the unknown side, x, while the 5-in. side corresponds to the 3-in. side in the smaller triangle.

Step 2: Set the corresponding sides up as a proportion.

$$\frac{9}{x} = \frac{5}{3}$$

Step 3: Cross multiply and solve.

$$9 \cdot 3 = 5x$$

$$x = \frac{27}{5} \text{ in.}$$

8.5 PRACTICE PROBLEMS

Solve the following proportions.

1. $\dfrac{5}{x} = \dfrac{4}{x+2}$

2. $\dfrac{2(x+1)}{4} = \dfrac{5}{2}$

3. $\dfrac{4x}{7} = \dfrac{2}{3}$

4. $\dfrac{3}{x+2} = \dfrac{4}{5}$

5. Solve for the unknown side.

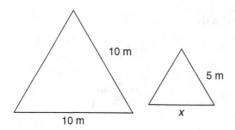

6. Solve for the unknown side.

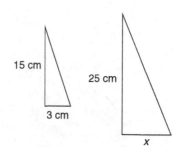